"Deeply disturbing…A vitally important new book."

—*Daily Mail*

"Unsettling…This story needs to be repeated and its grimmer details dwelt upon until it provokes action instead of unease."

—Mark Kohn, *Independent on Sunday*

"With a plot as strong as this, *The Estrogen Effect* could not fail to be gripping. She has researched her subject in extraordinary depth, and the result is a thorough, level-headed, and sophisticated book."

—Stephen Young, BBC

"A readable, concise, and accessible account…Her book reads like a detective story, as in fact it is: And it conveys the excitement of discovery very well…. [Readers] will be enthralled."

—Theodore Dalrymple, *Sunday Telegraph*

"A highly readable book with a dramatic presentation, this is a carefully crafted and footnoted history of a scientific debate that has profound implications for current public health and policy."

—*Library Journal*

The Estrogen Effect
How Chemical Pollution Is Threatening Our Survival

Deborah Cadbury

St. Martin's Griffin ⚐ New York

The author and publisher wish to thank the following for permission to reprint copyright material: Little, Brown and Co. (UK) and Penguin USA for extracts from *Our Stolen Future* by Dr. Theo Colburn, Diane Dumanoski, and Dr. John Peterson Myers; extracts from Chapters 2, 5, 6, 9, and 11 of Allan A. Jensen's *Chemical Contaminants in Human Milk* reprinted by permission of Lewis Publishers, an imprint of CRC Press, Boca Raton, Florida.

www.stmartins.com

ISBN 0-312-24396-0 (hc)
ISBN 0-312-26707-X (pbk)

First published in Great Britain under the title *The Feminization of Nature: Our Future at Risk* by Hamish Hamilton Ltd of the Penguin Group

First published in the United States under the title *Altering Eden: The Feminization of Nature*

First St. Martin's Griffin Edition: December 2000

10 9 8 7 6 5 4 3 2 1

CONTENTS

ACKNOWLEDGMENTS

It would not have been possible to write this book without the help of a great many scientists working at the forefront of the field who have contributed to my research over a number of years. All the scientists I met while making "Assault on the Male" for BBC *Horizon* in 1993 have continued to provide generous assistance. I would particularly like to thank Dr. Richard Sharpe, Professor John Sumpter, Professor John McLachlan, Professor Niels Skakkebaek, Professor Ana Soto, Dr. Theo Colborn and Professor Louis Guillette for their support in this latest project. I am grateful, too, to Jana Bennett, Head of BBC Science, who as editor of *Horizon* in 1993 helped me to develop these remarkable scientific ideas into a documentary. At the time none of the new data had been aired in the press and her judgment in evaluating its significance was invaluable.

As the field has grown, a great many scientists in Europe and America have contributed both to updating the *Horizon* film and to this book. Detailed interviews have been recorded with more than forty scientists and I am particularly indebted to those who gave up valuable time to read sections of the manuscript and offer expert advice. In addition to those mentioned above, I could not have completed the book without the help of the following specialists: Dr. Valerie Beral, Dr. Linda Birnbaum, Professor Frank Comhaire, Dr. Audrey Cummings, Dr. Philippa Darbre, Dr. Devra Lee Davis, Dr. William Davis, Professor Raymond Dils, Professor David Feldman, Professor Harry Fisch, Dr. Earl Gray, Dr. Timothy Gross, Dr. Stewart Irvine, Professor Harry Jellinck, Dr. Sue Jobling, Professor John Leatherland, Dr. Henrik Leffers, Gwynne Lyons, Dr. Peter Matthiessen, Professor Alan McNeilly, Dr. Fatima Olea, Dr.

Nicolas Olea, Dr. Michael Osborne, Dr. Ewa Rajpert-De Meyts, Dr. Mikael Roerth, Professor Frederick vom Saal, Professor Stephen Safe, Professor Risto Santti, Professor Carlos Sonnenschein, Dr. Charles Tyler and Dr. Mary Wolff.

As editor, Tony Lacey at Penguin provided encouragement and I am grateful for his skilled editorial judgment. Jane Bradish-Ellames at Curtis Brown helped me through the maze of the publishing world. Finally, special thanks are due to Julia Lilley for advice on the first draft and to Martin, Peter and Jo for their enthusiastic support.

FOREWORD

The world is in the grip of the ultimate epidemic. Infertility has spread like the plague. The ending has come with dramatic suddenness. Almost overnight, it seems, the human race has lost its ability to breed. After a global search, biologists have confirmed, nowhere in the world is there a pregnant woman. For 25 years no one has heard the cry of a newborn on the planet. The voices of children can only be heard on archive tape or television. Children's playgrounds have been dismantled. Schools have become centers of adult education. All over the world Nation States are storing their records for posterity, in some vain hope that some form of intelligent life will follow us. Without hope of a future for our species, each spring brings us closer to an ending no one could have foreseen a generation ago. It's impossible to view the horse-chestnut trees in full bloom, the sunlight moving on stone walls, without pain. We know there will be centuries of springs to come, unseen by the human eye.

In this vision of the future depicted by the novelist P. D. James, in her book *The Children of Men,* she envisages that what might enrage and demoralize us the most is not the impending end of our species, not even our inability to prevent it, but our failure to discover the *cause.* "Western medicine and Western science," she wrote, "have not prepared us for the magnitude and humiliation of this ultimate failure . . . Western science has been our god." It is hard to believe that it wouldn't still provide for us: "the anaesthetic for the pain, the spare heart, the new lung, the antibiotic, the moving wheels, the moving pictures. Now we face the universal disillusionment of those whose god has died."[1]

Such a scenario from science fiction may seem beyond belief now. Yet it was recent strands of scientific evidence that provided the background for the novel. This book examines that evidence and follows the true scientific detective story as some of the world's leading fertility experts uncover adverse changes to human reproduction. A dramatic fall in sperm counts has been reported. A Danish study found a 50 percent drop in sperm counts in fifty years; in Edinburgh, a 25 percent drop in the last twenty years; in Paris, a similar decline. Although some of the data has been challenged, other studies confirm worrying changes to human reproduction. There has also been a startling increase in testicular cancer and sex organ abnormalities of baby boys. Some scientists argue that if the decline continues at the same rate, it will not be too long before human reproduction will be under threat. This change to human reproduction is mirrored by extraordinary changes to wildlife. There are species showing signs of "feminization," their phalluses severely reduced in size, and others are mysteriously "changing sex," the males producing eggs just like females. Cancer, too, has been brought into the debate: the increased incidence of breast cancer and prostate cancer. It has been argued that all these changes have one thing in common. They can all be affected by exposure to the female hormone estrogen.

Recently it has been discovered that many man-made chemicals can act like "weak estrogens," imitating the female hormone and other hormones. Hormones are the most potent chemical messengers in the body because they act directly on the genes, instructing our cells how to behave and controlling critical body functions. There is evidence that some chemicals used in plastics, pesticides and many industrial products can mimic them, and may randomly create havoc with our reproduction and sex development, and may even play a part in some cancers. More frightening still, these are chemicals which we are eating, drinking, breathing and bathing in. They are chemicals which no human infant escapes, sometimes even from before birth.

Could these be worrying clues to our future? Or in the words of the chemical industry, is this no more than a collection of data which "generally fails to confirm the claim that industrial chemicals are causing widespread harm to health or the environment by hormonally mediated mechanisms?"[2] Some scientists point out that the field is fraught with uncertainties and there is insufficient data available to confirm what is causing these changes to man. They argue that certain studies even show that sperm counts are not falling and that scientists have yet to agree on this. Without hard evidence, they argue, it would be irresponsible to restrict or ban the use of a large number of man-made chemicals which are used to make many of the products that we need and take for granted. The costs are too great.

But for others the cost of not taking action could be even higher and we ignore the growing body of evidence at our peril. "Imagine if for the last fifty years we had sprayed the whole earth with a nerve gas which had very potent effects. Well, we've done that, not with a nerve gas, but with chemicals which act like hormone disrupters. We've released chemicals throughout the world that are having fundamental effects on the reproductive system and the immune system in wildlife and in humans. Should we change policy? Should we be upset? Yes, I think we should be fundamentally upset. I think we should be screaming in the streets," says Professor Louis Guillette of the University of Florida.

"We have unwittingly entered the ultimate Faustian bargain," argues Dr. Devra Lee Davis, former deputy health policy adviser to the American government. "In return for all the benefits of our modern society, and all the amazing products of modern life, we have more testicular cancer and more breast cancer. We may also affect the ability of the species to reproduce. I don't accept that bargain. The stakes are too high here. We cannot afford to take a course of action that will affect the ability of the species to persevere."

This book is an attempt to unravel the complexities behind these differing views; to tell the true story of how the new scientific ideas were uncovered; and to assess their possible significance for human health and public policy.

Prologue

Florida, 1990

It was pitch black in the middle of nowhere. The damp clinging heat from the gulf made it difficult to move easily. The rushes in the Florida swamps were at their most overgrown, shoulder high, and treacherous—home for a prehistoric reptile which hid just below the surface of the dark water. This was alligator territory.

Professor Louis Guillette and his team met Allan Woodward of the Florida Game and Fish Commission. Guillette's shirt clung to his back. His legs and arms were scratched and bruised from previous midnight expeditions into the swamps. The mosquitoes were irritating but he scarcely noticed. His team from the University of Florida had been invited to the lakes to solve a curious puzzle. Local farmers had reported that the alligator population on Lake Apopka had crashed. No obvious reason for this could be found. Apopka was the largest lake in Florida. The water stretched out before them like a sheet of black metal as far as the eye could see, until it merged with the night. As far as anyone knew, everything was the same as it had always been.

There were three airboats moored on the tiny jetty by the boathouse. They worked quickly, loading supplies and putting on life jackets. Each person wore a helmet with a strong lamp at the front, like a miner's. From a distance they made a strange sight, moving shafts of light in the dark.

There was a deafening roar as Woodward started the engine of the first boat. The blades cut into the black night and the boat skimmed across the rushes and sped down the canal toward the open lake. It was exhilarating. Louis Guillette and his colleague Dr. Timothy Gross were perched three feet above the water on seats at the front. Even though the breeze was hot, it was a welcome relief from the oppressive gulf air on land.

They knew the mission was risky. The aim was to catch the juvenile alligators with their bare hands and take blood samples to see if this gave any clues as to why the population was in decline. Wild alligators have to be caught at night when they are most active. The only clue to their presence are the red eyes glinting on the surface of the black water, two red points barely decipherable in the night. One misjudgment and you could suffer a severe bite or worse. There was also the danger of falling into the water with the alligators. Only the previous summer an airboat had rolled. Guillette had spent the night on a floating island in the dark, waiting for help the next morning.

They navigated for ten minutes in silence across the lake. Louis Guillette knew where the adult alligators were likely to be. By day his team had already studied the nests and they knew they had stumbled upon a serious problem. Over 75 percent of the eggs they studied were dead. Yet the lake was classified as environmentally clean; the water was supposedly free of any chemical contamination. It was hard to see why the eggs were dying.

As they reached the far side of the lake and approached the rushes, they switched off the engine. The sudden silence was deafening. The boat rocked gently. The flashlights of their helmets made curious shafts of light on the water. Suddenly they both saw two dots of red, shining in the flashlight. Then another pair, and another. The airboat was surrounded by a group of alligators scarcely twenty yards away.

"We've got a six-footer . . . Over."

"Can you move in fifty meters to your left."

The talk back from the other airboat on their headsets broke through the silence. They could see several pairs of eyes still watching them. The actions that followed took just seconds and seemed effortless. As Woodward restarted the engine and swung the boat toward the nearest alligator, Professor Guillette leaned well over the front of the airboat, his arms outstretched into the darkness. He knew you only got one chance. If you missed, you had to let the animal go. It was too risky to try again. In a split second, he had grabbed the alligator behind the neck, the only place to hold it safely. Holding the limbs or tail is fatal because they are free to run around and bite. The alligator's teeth, up to three-quarters of an inch in a six-foot animal, could inflict a serious wound. Infection and tearing were the greatest dangers.

But this alligator scarcely moved once caught in their deft movements. Professor Guillette held the jaw shut as Dr. Gross pulled the body up onto the boat. Soon the vast jaw, shaped in a disconcerting half-smile, was taped shut. The scales were thick and creased; the yellow, black and green markings down the spine were pronounced. The odor was strange, reptilian. The animal from prehistory was about to meet modern science.

In a matter of seconds the huge animal was restrained on the side of the boat. They took a blood sample effortlessly and studied the creature for any obvious defects. As they turned the alligator and studied the reproductive organs, it was clear immediately that this male was not going to be able to reproduce. It was typical of many of the animals they saw that night.

"We were astonished at what we found," recalls Professor Guillette. "Things we were seeing were so dramatic. We were actually seeing sex reversal. The alligators seemed to be *changing sex*. The animals were neither male nor female. They were partially sex reversed. Up to 80 percent of the male alligators we've studied on this lake have some kind of abnormal penis. Mostly the abnormality is small size, sometimes as much as a half or two-thirds reduced."[1]

There was a quiet splash as the animal was returned to the water. It slid under the black surface effortlessly and vanished.

Night after night Professor Louis Guillette and his team went to the lake to make their observations. The pattern was repeated endlessly. Many of the adults were infertile. Many of the eggs that were produced were dead. Yet the water in the lake was supposed to be clean. When water samples were tested, there was no obvious major contaminant. So what was causing the problem?

Back in the lab, under the microscope they could see mysterious alterations to the structure of the testes so that the creature no longer looked like a normal male. "We never set out to study sex reversal or environmental contamination," explains Louis Guillette. "We just wanted to know what was going on in Lake Apopka. But the things we were seeing were so incredibly different that it just hits you in the head. There was something fundamentally different about this lake which demanded that we come back to study it. The data from the animals was so strange."[2]

When they analyzed the blood samples, they finally had at least part of the answer: the alligators had high levels of the female hormone estrogen. In many cases the levels were double the normal concentrations. Male alligators also had depressed levels of the male hormone testosterone. Somehow, hormone levels vital for reproductive health had been scrambled.[3]

Unknown to Professor Louis Guillette, at least six other scientific teams in different parts of the world were identifying similar changes to the males of other species, including man. In London, scientists were finding fish "changing sex" in British rivers.[4] In Paris, "feminized" eels were found on the Seine.[5] In Florida, panthers were discovered with severe abnormalities of the reproductive tract.[6] In the Great Lakes, several species were found with similar disorders, failing to reach sexual maturity and other fertility problems.[7] Guillette's work was to provide a compelling piece of the jigsaw in the chilling picture that was about to emerge. Professor Louis Guillette concluded from his wildlife studies that "everything we are seeing in wildlife has an implication for humans. I believe that we have the potential to have major human reproduction problems."[8]

Edinburgh, Scotland, February 1996

Dr. Stewart Irvine, a clinical consultant at the Medical Research Council in Edinburgh, was surprised when he heard of the first reports of a dramatic fall in human sperm counts in 1992. "That bold figure, a 50 percent drop in fifty years worldwide, was very striking, especially after spending many years dealing with infertility and knowing how variable sperm quality can be. I was quite skeptical," he recalls.

As a consultant gynecologist in the fertility clinic in the Royal Edinburgh Infirmary, Dr. Irvine had been dealing with infertile couples for many years. All too often, in nearly half the cases, the difficulties in starting a family lay with the man, not the woman. "Men don't expect to be told that they have a fertility problem," he says, "it usually comes as an immense shock." Even with years of experience, and a highly skilled team, there was, quite simply, no easy way of breaking the news. Some people would be aggressive, some heartbroken. More often, people simply couldn't take in what was being said and would just look blankly, unable even to ask questions. "The infertility clinics are emotionally very difficult clinics to do. You're dealing with couples who are very, very distressed at their failure to conceive. I don't think, as a doctor, that you ever become immune to the emotional pain. The day I stop being emotionally affected by what is happening to my patients will be the day I stop practicing medicine."

Undoubtedly the clinic had been very busy. But a 50 percent drop in sperm counts in fifty years stretched credibility to the limit. Dr. Irvine knew he had data and semen samples from men over the last twelve years, which had been collected for research purposes. He was confident that this data would quickly reveal that in Scotland, at least, all was well.

If it was true that sperm counts were falling, he reasoned, then younger men, in their twenties, should have much lower sperm counts than men who are now in their forties and fifties. This seemed highly unlikely. Nonetheless, they set out to analyze whether there was any difference in sperm counts in four age groups: men born in the 1950s, early 1960s, late 1960s and early 1970s. His team analyzed data on over 500 men, aged between 18 and 45, using the same sperm-counting techniques throughout the study.

Human sperm can be easily seen under the microscope. At first sight there did indeed appear to be a problem. He found a typical sample from a man born in the 1950s contained many tadpole-like sperm swimming quite vigorously across the field of view. Although there were some abnormal sperm, by all the conventional criteria of shape, movement and volume the vast majority of the sperm appeared normal. However, for a younger man, born in the 1970s, what he saw down the microscope was very different. There were fewer sperm in the frame of view. Some of these were deformed. Often the motility of the sperm was very poor, with misshapen sperm maneuvering themselves clumsily across the field of view. Several of the sperm didn't move at all. These deformed and cumbersome tadpole-like structures, the genetic essence of maleness in a single cell, held little promise as the start of the next generation.

In his office at the Medical Research Council, Dr. Irvine analyzed the data. It was not long before a pattern began to emerge. Men born in the 1950s had an average sperm count of about 100 million sperm per milliliter of semen. Men born in the early part of the 1970s had an average sperm count of 78 million per milliliter. It was a drop of about 25 percent in just twenty years. For a change in a physiological process, this is staggeringly fast. But with typical scientific caution, his response to these extraordinary results emerging on the computer screen was unemotional. "I didn't look at the data on the screen and think, oh my goodness, how terribly exciting! I don't think science is like that. There was no moment when it suddenly became clear: a 25 percent drop in twenty years. I thought, this is an interesting result. This is a little bit of evidence in a big jigsaw puzzle." He discussed it with his colleagues John Aitken and Elizabeth Cawood and scrutinized it for any flaw.

He began to prepare an article on his results for the *British Medical Journal*, sometimes working in early evenings with his young son, who was waiting for a turn on the computer. To anyone walking by along the corridors of the Medical Research Council, they would have glimpsed a homely scene of father and son together, heads bowed apparently engrossed in a computer game. The small office was always tidy. The white lab coat hung by the door. On shelves above the desk, years and years of study were carefully filed away. The noticeboard on the wall contained the important details, the timings of the fertility clinics, forthcoming conferences, and so on. Everything was as it had always been: part of a busy, humdrum and well-ordered routine. But not for long. This was the kind of result that would disrupt everything.

Every way he plotted the data it was a straight downward slope from the 1950s to the 1970s. It was hard to believe, even though it was his own data. Even though he saw this as just a little bit of data in a big jigsaw, that jigsaw was baffling. This was, after all, the process of creation that was in question. Yet he was certain his own data was correct. He knew it tallied with data from at least three other studies. Still it seemed extraordinary that we might have interfered with the process of creation in such a fundamental way, a process which nature had ensured was so abundant and prolific that it seemed beyond our interference.

Soon after sending his paper to the *British Medical Journal,*[9] he received a call from them. "We're planning a press conference. Would you be willing to participate?" He was taken by surprise, but agreed.

"Dealing with the press was just awful," he recalled later. "I never want to do it again. The worst day involved twenty-three interviews. I found the whole episode very difficult, because the press are not interested in the fact that you have reservations about your own data. They just want it talked up so that it is a newsworthy story. I remember having a conversation with one TV journalist who asked if I were to appear on the program would I take such and such a view, and then presented a very extreme scenario. I said no, I don't think that is a rational point of view. And she said, oh well that's not newsworthy, then, we'll get somebody else."

Nonetheless, despite his concerns, there were serious questions raised by his data. He explained them to *Horizon.* "A drop of this magnitude was really quite surprising. I was concerned," he told us. "What's rather more alarming is that our data suggests these changes are continuing. The youngest group we had in the study was born in the 1970s and we simply don't know what is happening to boys who are being born now. Who knows what the semen quality of boys who have been born in the 1980s and early 1990s is going to be in twenty or thirty years' time when they grow up. . . . Now clearly *if* this decline is continuing—and our data suggests that it is—and it carries on down, then men who are being born now, in the early part of the 1990s, are going to grow up and have a sperm count of about 50 million per milliliter. That's worrying I think, because in an infertility clinic context we tell a man that his fertility is compromised when his sperm count falls below 20 million per milliliter. *This makes me, at least, look at this data and stop and be worried, frankly, that this is a substantial change that appears to be continuing. Clearly one of the anxieties that I have is that we just can't afford to wait twenty or thirty years until this generation grows up to find out that, yes, things have got worse and, no, we've done nothing about it. I think it is really urgent that work is done now to address whether those changes are continuing and, if so, what can be done about it.*"[10]

Washington, D.C., February 1996

At the same time that Dr. Stewart Irvine was preparing to meet the press with his results, representatives of some of the wealthiest and most powerful industries in America, headed by the Chemical Manufacturers Association, met to form a special task group. Their aim: to work out a coordinated strategy to a growing number of worrying reports in the media and elsewhere suggesting that some man-made chemicals used in plastics, pesticides and industry could be affecting the reproduction of humans and wildlife.[11]

In a summary of their findings, the Endocrine Issues Task Group noted that the "impact on the chemical industry could be *far reaching.* Several specific substances and groups of compounds have already been implicated . . . these issues could also affect industry sectors outside of chemical manufacturing." Some customers were already deselecting certain chemicals.

The task group found: "At a time when Congress and the public are becoming more receptive to science-based regulation and cost-benefit decision making, widespread concern about endocrine effects could spur a return to conservative regulatory models under which exposure to chemicals is controlled to the greatest degree possible in order to compensate for scientific uncertainties."

The minutes also analyzed the media interest in the topic including BBC programs: "Television news features such as 'Assault on the Male' . . . illustrate the powerful link between concern about the alleged hormonal effects of chemicals and widespread public anxiety about perceived declines in reproductive performance as well as increases in breast cancer and other emotional public health concerns."

It was agreed that there was a "need to allay public concern." Among the many objectives in their "Action Plans" they noted that they should:

- enlist "scientific opinion leaders' from different disciplines, including the medical profession, to help *focus* media coverage and public debate on sound science rather than speculation.
- develop and implement plans to work closely with Congress, the Environmental Protection Agency, state and federal agencies, the media, and the public on endocrine issues to take action where justified but avoid premature or unnecessary regulation/legislation . . .
- develop an effective education program . . .
- examine the existing data base to determine the level of scientific support for key elements of the endocrine hypothesis.

It was also agreed that "forming global partnerships should also be a high priority. US Industry should interact with the appropriate groups in the EC [now EU] through regular meetings/video conferences."[12] A coordinated response by the global chemical industry was needed.

The conclusions of the task group were carefully typed out and the report circulated to the Environment, Health, Safety and Operations Committee.

At the same time as the chemical industry was working out its strategy for dealing with these allegations, evidence was also being gathered by Congress in Washington. The American National Academy of Sciences had launched a major study of the chemicals and the Environmental Protection Agency was set to make them a top research priority. Even the Vice-President of America himself had written an impassioned account of the issue, foreseeing a future troubled by "infertility; genital deformities; hormonally triggered human cancers, such as those of the breast and prostate gland; and neurological disorders."[13] In Europe, special committees had been set up to investigate and in Whitehall in London, true to style, officials had classified key research, a move they would later have cause to regret. On both sides of the Atlantic, a scientific debate which no one had heard of four years ago had exploded into a major industrial, political and public health issue. The battle lines in an extraordinary scientific debate, which some believe might affect the survival of our species, were being set out.

This is the true story of what those scientific discoveries were and how they were made.

CHAPTER ONE

Early Warnings

Copenhagen, Denmark, 1991

While many scientists may dream of making a significant discovery, very few succeed. That Niels Skakkebaek should have achieved this before he had even collected his Ph.D. might well have aroused jealousy in others in the field, were it not that, according to his colleagues, he also happens to be extremely likable and dedicated to science. "He's very aware of connections," explains Dr. Henrik Leffers from Copenhagen University Hospital. "If I suggest an idea or hypothesis then he very rapidly thinks three or four steps further on." "He's very tenacious, and very creative," says Professor Mikael Roerth. "He's also got an inborn authority: he is a natural leader."

Now at the height of his career, Niels Skakkebaek has worked as an adviser both to the Danish government and to the World Health Organization, and leads the very busy department of Growth and Reproduction which he created at Copenhagen University Hospital. Here, his team sees thousands of young patients a year, specializing in male infertility and hormone disorders affecting reproduction, delayed puberty, precocious puberty and intersex conditions. More than this, the research from the department has attracted worldwide attention and inspired scientific teams. "The science itself is his first priority," says Dr. Ewa Rajpert-De Meyts. "If he can, he will just drop everything and come to the laboratory with you if there is an interesting result. He's not very easily impressed. It needs to be very carefully and rigorously done to satisfy

him that what we are seeing is true, although he is always enthusiastic."

This complete dedication to scientific research has led Professor
Niels Skakkebaek over thirty-five years to one particularly worrying
conclusion. There is a serious problem with male reproductive
health. What is more there are signs that at least some of the adverse
changes are increasing. "We see the findings as the writing on the
wall," warns Niels Skakkebaek. "Something is affecting human reproduction and the frightening thing is that we do not know the
cause. It may well have implications for health in general. We see
this as an early indicator that something is wrong." Faced with the
controversy that some aspects of his view has created and the uncertainty over how to act, he replies with his serious tone: "You
can only raise a flag. The writing is on the wall that we face a
problem. But I think I've done enough. I've told the world there
is a problem. I'm not skilled politically to say what you should do—
that is very complicated. But as a physician, dealing with human
beings, I see all these problems. I think it is dreadful. The number
of young boys with testicular cancer. The number of women with
breast cancer. It is all too much."[1]

He stumbled on the first clues that there was a problem with
male reproductive health over twenty-five years ago. While doing
his doctoral thesis, Niels Skakkebaek was trying to investigate the
causes of infertility. This involved examining tissue samples under
the microscope from over two hundred infertile men. It was typical
scientific research, painstaking, pedestrian, with rigorous attention
to detail. But one day, he came across some strange cells in the
testes, quite different from the others. These cells had not been
identified and classified before. The nucleus, which contains the
genetic material, was coarse and irregularly shaped, packed with far
too much DNA. The rest of the cell was unusually empty, with
hardly any structures in it. This was not what he expected to see
at all. "I thought immediately these cells were abnormal and of
significance," he recalls, "although I did not know the meaning of
it. But it was quite clear to me that they had some excitement in
them. It was also quite irritating that they did not fit my system for
classification. I had no idea what they were." He had studied with

Professor Yves Clermont at McGill University in Montreal, who at the time was an acclaimed authority on the classification of cells in the testes and had taught Niels Skakkebaek how to examine tissues.

But the experts, the pathologists who specialize in studying cells under the microscope, were most dismissive. These strange cells were nothing, of no significance whatsoever. Skakkebaek was not so sure. Rather than jettisoning the material, he stored the mysterious cells away. By chance, two years later, he came across exactly the same type of cells in another infertile patient. The original samples were brought out again to compare. There was nothing in the literature about such cells in infertile men. "It struck me that it must have some meaning that the two patients with exactly the same problem turn up with exactly the same abnormal cells," remarked Skakkebaek. While he was preparing a paper about these new cells, there was an unexpected development. It was a day which he still remembers vividly.

The first patient came back to the department for further investigations to help with his infertility. A doctor took a biopsy and fetched Niels Skakkebaek immediately. There was something very, very wrong. The patient had developed a cancer. It was in the testes, right on the very site where the strange cells had been. They sent the results to the laboratory for an immediate check. The results were back within half an hour. This was unquestionably a tumor. What is more, it was a very invasive and fast-growing tumor.

Niels Skakkebaek, who at this stage knew very little about testicular cancer, had never before told a patient that he had cancer. To make matters worse, this was a young man embarking on his career, who thought he had all his life ahead of him. The events of the next hour made a lasting impression on Niels Skakkebaek: the small closed room, the intensity of the discussion, the anxieties of the patient. He explained the results and made arrangements for surgery and further treatment.

Afterward, Skakkebaek went over events in his mind. It struck him with force. The strange cells he had seen in this patient a few years earlier could well be the precursors of testicular cancer. These cells had somehow developed into the tumor. He discussed this with his supervisor and went on to propose this in his doctoral thesis. Confident he was right, he even went so far as to name the new cells "CIS testes," which means carcinoma *in situ*.[2]

However, the medical profession was not that easily impressed. Who, after all, was this young upstart, not even out of graduate school, who thought he could solve critical problems in cancer diagnosis with just a few sharp observations? His articles were rejected by several leading scientific journals, rejection letters which he still has to this day. Only the British journal, the *Lancet,* was interested enough to publish his first paper on the precursors of testicular cancer. It sparked immediate controversy. Pathologists, especially in America, who were the experts in the classification of diseased cells, were adamant that he was wrong. "This really kicked against the way of thinking," explains Skakkebaek. "It was particularly provocative because I was not a pathologist. I was unknown, I was nothing." Yet here he was daring to tell pathologists that they had missed something important.

Skakkebaek was unshaken by the controversy. Arguments raged over whether these cells were produced by the tumor, or caused the tumor, and how such cells could produce all the different types of testes cancer. "I knew they were wrong, I was completely certain. I did not have a moment's doubt in my mind because I had seen the patients." By now he had seen eight patients who had these cells and went on to develop cancer. "The arguments used against my idea were not good. They were very superficial. You know, this is not part of our usual way of thinking. That is no argument. . . . All good research challenges existing knowledge. The aim of research is to uncover biological relationships and it is inevitable that this will involve challenging preconceived notions." Eventually, other fertility experts began to see these cells in patients who went on to develop testicular cancer. It took ten years, but his ideas are now almost universally accepted. His discovery has led to much earlier diagnosis of the disease and improved treatment.

As Skakkebaek continued his work on testicular cancer, he was surprised at how often he was seeing patients with this disease, which was supposed to be exceptionally rare. He searched the cancer registries, and uncovered a disturbing trend. The incidence of this cancer had risen sharply, by three- or fourfold in the last forty to fifty years.[3] "In Denmark we now have 300 percent more testicular cancer than we had fifty years ago," Skakkebaek explains. "We actually have the highest incidence in the world, and the frightening thing is that we really do not know why." Wherever he looked, he found the same trend: Britain, the Nordic and Baltic countries, Australia, New Zealand, and America.[4] The average increase was 2 to 4 percent per year over several decades, making testicular cancer the commonest cancer in young men.[5] He knew the data from cancer registries was very reliable. Testicular cancer is not something that can be misdiagnosed to any great extent and the change in incidence could not be explained by altered diagnosis. This increased incidence was a real and worrying trend.

The disease can be all the more distressing because it affects young men, which is unusual for cancer. The tumor is most likely to be diagnosed during or after puberty, when the hormonal environment changes. Adrian Hill's case is typical. Adrian Hill was diagnosed when he was just twenty-five years old at a hospital in Edinburgh, where he lives.[6] "My wife first noticed there was something wrong. One side was larger and harder than the other. I tried to just ignore it to begin with, thinking there was nothing wrong and left it for a while. Then one day I came home from work and could hardly move. I began to cry with the pain. My wife phoned the doctor and we were prescribed antibiotics for a fortnight. But I was no better and went back to the doctor and eventually was referred to a specialist. By now some weeks had passed since my wife first noticed the changes."

Such difficulty in diagnosis is typical. Dr. Ewa Rajpert-De Meyts from Copenhagen University Hospital explains: "For a long time the tumor does not give any special symptoms. There can be a bit of enlargement of the testicle which may worry young men, just a little swelling, usually without pain. However, the problem is that the pre-invasive stage of the disease can sit inside the testes for many many years. It sits there dormant until puberty when there is a big change in hormonal milieu and then it wakens up and develops into cancer. The more invasive types may spread, or metastasize, before producing any symptoms in the testes and then the first signs may be pains in the abdomen or the back, depending on where it has spread."

"Spread is nearly always to the lymph nodes between the kidneys, and when they get swollen this will give back pain," says Professor Mikael Roerth, a cancer specialist at Copenhagen University Hospital. "If it has spread further it is often to the lungs, and that will give you a cough; sometimes you may cough up a little blood. These patients are for the most part looking completely normal, feeling fit and everything."

Adrian Hill was told he needed an operation to remove one testicle and signed the consent forms. "Once you hear the word 'cancer' it feels like a death warrant. It was very frightening. My wife was pregnant at the time and I was worried I was going to be leaving them both behind. I just kept thinking: Am I going to die?"

The treatment for a young man is always as sparing as possible, to try to preserve fertility. Before treatment, Adrian Hill was advised to donate sperm to a sperm bank since any radiation or chemotherapy could well affect his future fertility. "I'd always wanted another child and I know my wife does, so the treatment was a very big step," he recalls.

"In most cases the cancer occurs on one side only, but there are a few cases where it occurs on both sides and then there is no way you can spare fertility because both testicles will have to be removed," says Dr. Ewa Rajpert-De Meyts. "A number of tests can be done to find out how far the cancer has spread and then we treat this with aggressive chemotherapy and sometimes also irradiation." In Adrian Hill's case, the medical team found the cancer had not spread. He stands an excellent chance of making a complete recovery.

"Today the realistic goal of every patient with testicular cancer is cure," says Professor Mikael Roerth. Survival rates are high, in the range 90 to 95 percent. "This is because the cells in the testes, known as germ cells, are extremely sensitive to radiation and chemotherapy and they just melt away," explains Dr. Ewa Rajpert-De Meyts. "Consequently tumors derived from these germ cells, even if they have spread beyond the testes, respond to treatment much faster than some other types of cancer cells. This extraordinary sensitivity of cells in the testes to damage is believed to be nature's way of ensuring that if there is any genetic damage to cells that will produce the next generation, these cells self-destruct. It protects the species against damaging mutations." It is this unusual sensitivity of cells in the testes to chemicals, radiation and other environmental factors which makes them a good marker for human health trends in general.

Skakkebaek was interested in why this distressing cancer should be increasing. Casting his mind back to his early work, he remembered something else remarkable about the strange cells that serve as precursors to the disease. "These cells look much more like very primitive cells that you see in embryos. You don't expect to find in an adult man cells that look like the cells of a fetus. It was quite a striking observation." It suggested that the events that led to testicular cancer were perhaps occurring very early in life, in the womb, when the cells may become fixed prematurely at an early stage of development. Further studies confirmed these cells could indeed be found in some aborted fetuses, lending weight to the idea that the events leading to testicular cancer occurred early in development. Skakkebaek hypothesized that testicular cancer in the adult is caused by some prenatal event, although the disease does not develop into its pathological state until after being stimulated by hormones at puberty. His ideas are now widely accepted. This work provided the first important clue that something might be affecting the development of baby boys in the womb.[7]

But there were other clues suggesting male reproduction was being damaged. By the mid-1980s, Skakkebaek's studies on testicular cancer had gained him a considerable reputation. He found the research very satisfying. "Sometimes it takes many years to find out what makes good science," he says, "just as it can in the arts. But if it is good science the creative process is beautiful. There was nothing there before, and suddenly it is there—a line of thought you have uncovered, and it is lasting. The creative process is the same in science as in the arts. You start with a stone and build a sculpture. There was nothing there, and you build something. That is the beauty, if it is solid and good."

He continued his research in Seattle and Edinburgh, building up yet more expertise in male infertility and endocrinology, the study of hormones. But he still dealt directly with patients, over a thousand a year, in his clinical practice, which he felt was important to steer his research. Eventually he was appointed head of the department of pediatrics in the busy Hwidovre Hospital in Copenhagen and built up a team of experts dealing with hormone disorders in children. He was also running research and a clinic at the Copenhagen University Hospital. His work with patients led him to suspect that testicular cancer may not be the only distressing reproductive disorder in baby boys showing an increase.

The process of development of the fetus in the womb is one of the marvels of nature, a process so complex biologists have scarcely begun to unravel the biological relationships, the cascade of events that needs to occur to ensure the organs form correctly and that they end up in the right place in the body. The development of the reproductive tract is no exception. A whole series of hormonal cues have to occur at the right time to orchestrate events that lead to the development of male or female. The human fetus always starts life as a female. However, if it is to become a boy, at around six weeks, the male sex glands, the testes, start to differentiate, one on each side of the body, close to the kidneys. In later life, these will produce the sperm, but even in the fetus they start to secrete the male hormone, testosterone. This interacts with other hormones to trigger a series of events leading to the development of the male reproductive tract. The female ducts regress and are reabsorbed. At about ten weeks the testes usually start migration from the kidneys toward the scrotum, which is reached just before birth. When one or both testes fail to descend from a position near the kidneys into the scrotum, it is known as *testicular non descent*. Because testicular non descent in baby boys is associated with testicular cancer later in life and sometimes with reduced fertility, it is usually corrected surgically at around two years.

Skakkebaek found evidence that this condition, too, has increased and is now quite common, affecting 2 to 3 percent of all baby boys. Surveying the literature, he could see that concerns were first raised about an increase in testicular non descent back in 1984. Dr. Clair Chilvers studied hospital records for England and Wales and found the number of baby boys discharged from hospitals with this diagnosis had doubled from 1961 to 1981. The rate of incidence appeared to have increased from 1.4 to 2.9 percent in just twenty years.[8] Using a different set of data, Dr. Campbell at the Communicable Diseases Unit in Glasgow analyzed Scottish records on testicular non descent. He too found a dramatic increase, from 326 cases in 1961 to 2084 in 1985. "Surveillance of this trend is required," he warned. "If the incidence of undescended testes has truly increased, the reasons for this trend need to be ascertained."[9]

These reports prompted a letter to the *Lancet* from Valerie Beral at the Epidemiological Monitoring Unit, London. She and her team had studied national statistics on congenital malformations based on records of malformations noted at birth, which would avoid the bias of trends in hospital practice. These figures, too, showed a rise in the frequency of undescended testes from 1964 to 1983. In addition they found "rapid" increases in the incidence of other deformities of the male reproductive tract. "Could there be a common explanation of these trends?" she asks in her letter to the *Lancet*.[10]

Further studies were carried out which appeared to continue to support these trends. One study at the John Radcliffe Hospital in Oxford reported a staggering 40 percent increase in testicular non descent at birth compared with a similar study in the late 1950s.[11] Skakkebaek and his team found all this data was harder to interpret than the figures on testicular cancer. Sometimes it was difficult to tell from the hospital records whether the apparent increase was due to a change in surgical or diagnostic practice, or whether this represented a real increase in the condition. Nonetheless the best studies, which tried to allow for changes in practice, still suggested that there had been more than a doubling in the incidence of the condition in the last thirty to forty years. Because this is an abnormality present at birth, like testicular cancer, this data suggested that something might be going wrong during normal fetal development.

Their inquiries might have stopped there, were it not that there were other signs of adverse changes. They found that yet another distressing abnormality of the reproductive tract of young boys was showing an increase: a condition known as "hypospadia."[12] "Hypospadias are congenital abnormalities of the penis. It can have a severe form or a very mild form," explains Niels Skakkebaek. "In its mild form it is just a little cleft in the penis, where the tip of the urethra opens halfway up on the inside instead of at the tip. But in its more severe form it can be so extreme that you can really be in doubt as to whether this is a boy or a girl. In extreme cases it is in effect an intersex condition, with a mixture of sexual characteristics."

Sometimes the hermaphroditism involves not just the external genitalia, but also the internal organs. The testes and ovaries may be modified, forming an intergonad, which has characteristics of both a testis and an ovary. Frequently, in such cases, there can be abnormalities to the penis and the clitoris, such as a clefted scrotum, a reduced penis, or a urethra not opening at the tip of the penis. For the individuals concerned, the sex can be difficult to determine. Usually a series of operations will be performed to try to make a male or a female. Almost always the individual will be infertile.

Birth data for this condition from several reports have also indicated an increase in the prevalence of hypospadias over the last thirty to fifty years. In some countries, such as Denmark, Sweden and Norway, the increase has been 2 to 3 percent a year over a few years; in other countries the registers are less clear.[13]

It was beginning to look as though there might be a pattern. Testicular cancer, testicular non descent, hypospadias—all disorders whose origins in some way could be traced back to development and events early in life. Despite uncertainties over some of the data, Skakkebaek thought that this was too much to be a coincidence. There was also anecdotal evidence of a decline in quality of human sperm. He had been dealing with patients suffering from infertility all his career. Reports from his sperm laboratory suggested abnormalities were increasing.

He was still running two different hospital departments in Copenhagen and becoming increasingly frustrated that there was too little time for fundamental research. In 1990, he was able to amalgamate the two departments to create the Department of Growth and Reproduction at Copenhagen University Hospital and bring together the expertise of the two groups. This was a unique department, dealing with children's hormone disorders, growth and development, problems with puberty, male infertility, and intersex conditions. With more time now to tackle some basic research, the team really got going.

They conducted a preliminary study on healthy young men in their clinic. The results were startling. Many of the semen sample contained too few sperm and many of the young men had more than 85 percent abnormal sperm. Skakkebaek knew of studies twenty years previously claiming a decline in human sperm counts. But these studies had all been ignored or "explained away" on the grounds of inadequate data. He began to wonder whether there might have been something in them after all.

In the early seventies two doctors, Kinloch Nelson and Raymond Bunge, at the University of Iowa Hospital in America, had become concerned about an increase in the number of men coming to them for infertility treatment. They carried out a study of nearly four hundred men who had completed their families and were planning to have a vasectomy. To their surprise, they found that this group of apparently normal fertile men had average sperm concentrations of only 48 million per milliliter.[14] This was literally half the expected concentration. Previous studies had established that the average American male had sperm counts of well over 100 million sperm per milliliter. One study in 1949 had even found the average sperm counts to be 145 million per milliliter.[15]

"Dr. Kinloch Nelson was very highly respected," recalls Niels Skakkebaek. "I had no reason to believe that they were wrong. But their work had been challenged by John Macleod who was considered *the* great spermatologist of the time. He had the opinion that they were just wrong."

Indeed, Professor John Macleod of Cornell University in New York was so highly regarded he was known to some as "the king of spermatology." In a monumental study of 2,000 men in the 1950s, he had established the conventional criteria by which human sperm should be judged. In this study, over 44 percent of men had sperm counts over 100 million.[16] This was in stark contrast to Nelson and Bunge's study in the 1970s, which found only 7 percent of men had sperm counts above 100 million.

Perusing the studies, Skakkebaek could see that Nelson and Bunge were so concerned that their results were startlingly different from the earlier studies that they went back and checked the records of their 1950s patients at the Iowa University Hospital. These men did indeed have much higher sperm counts. Over 25 percent had concentrations greater than 100 million per milliliter. The doctors searched for any errors in their techniques and brought in an independent pathologist to assess their counting methods. It made no difference. "This tended to confirm our suspicions that something had changed in the intervening years since Macleod's study," they wrote ominously. "Something has altered the fertile male population to depress semen analysis remarkably. . . . The overall decrease in sperm concentration would tend to incriminate an environmental factor to which the entire population has been exposed."[17]

This was a direct challenge to Professor John Macleod, who was by now a very senior figure in the field of male fertility. The Iowa study suggested that the standards he had set no longer held true. What is more, two further studies were published also reporting a decline in sperm counts and semen quality.[18] It was not long before John Macleod published a detailed critique of the Iowa study. "We could not and cannot reconcile their astonishingly low value of only 7 percent having a sperm count above 100 million per milliliter with *any* population appearing in our laboratory at *any* time," Macleod declared. He disputed the idea of a larger overall decline and suggested that Nelson and Bunge's findings might be simply the result of analytical errors[19] Such was Macleod's standing in the field that the subject of declining sperm counts was closed, at least for the time being. Other studies raising a warning, such as those of Dr. Salvatore Leto and Dr. Frederick Frensilli in Washington, who found a marked decline in sperm count and sperm quality, were published. But such studies passed without comment.[20]

Niels Skakkebaek was aware of the controversy and had always been impressed with Macleod, so he was skeptical of a decline in sperm counts. Nonetheless, spurred on by his preliminary findings he set out on a much more comprehensive search to try to find out if sperm counts really were changing worldwide. "I did not feel any apprehension in reopening the debate. I just felt curious. I would never have done it if I didn't feel there was a problem. But male infertility *is* a problem," says Skakkebaek.

They soon found that data on sperm counts had been published in the international scientific literature since 1938. Dr. Elisabeth Carlsen, a research fellow getting her doctorate, was brought in to coordinate the study. With the help of the archivists at Copenhagen University library, she tracked down all the reports of the last fifty years on sperm counts from around the world. Apart from the United States, there were studies from Europe, South America, India and the Far East. They contacted Professor Niels Keiding, an eminent statistician at the Panum Institute in Copenhagen to discuss the design of their study. To exclude any risk of selection bias from their analysis, they discounted any research where men were sampled at fertility clinics, such men being liable to particularly low sperm counts, and also any studies where sperm was counted in a different way. In the final statistical analysis they included over sixty studies in the last fifty years, covering a group of almost 15,000 men.

The results were far more clear-cut and far more disturbing than any of them had anticipated. "We were very surprised that we found a dramatic fall in semen quality," says Niels Skakkebaek. "The number of sperm per ejaculate had fallen by approximately 50 percent during the fifty years. We also found a significant decrease in the volume of the ejaculate." The average sperm count had declined dramatically from the 113 million per milliliter of semen in 1940 to 66 million per in milliliter 1990. They also found a threefold increase in the number of men whose sperm count was below 20 million, the level at which their fertility would be affected.[21]

Their first thought was that this must be due to some systematic error in their research. They searched the data for any flaws in the statistics or methods of analysis. But no matter how they examined the data, worldwide or America alone, they found the same alarming trend. They started a systematic search for any alternative explanations for their findings. Were there, for instance, any changes in sperm-counting techniques during the last fifty years? But this too seemed unlikely. The samples once washed and prepared are placed under the microscope and counted manually with a desktop counter. It's a laborious technique, which has remained much the same since the 1930s. They checked the data on the other cells in the body, such as blood cells. These are counted in the same way, yet there was no change. People have the same number of blood cells as they did fifty years ago. They even checked out changes in sexual habits. It was possible these had changed during the last fifty years and this could affect sperm count. Once again, there was no evidence that this was the case.

Although sperm counts appeared to be declining, there was much less data on male fertility. A 50 percent drop in sperm counts does not correspond to a similar decline in human fertility. This is because there is no simple relationship between the number of sperm produced and whether or not a man is infertile.

Whereas a woman produces only one egg a month, men produce sperm in astronomical numbers, around 100 million a day, a thousand per second. Although it would appear with such abundance that there was little to worry about, in practice things are not so straightforward. "There is likely to be some difficulty if sperm counts drop below 30 million per milliliter," explains Skakkebaek. Most experts agree a man will have difficulty starting a family if sperm counts fall below 20 million. Below 5 million, the man is often sterile.

It is thought that the reason for this superabundance of sperm is because of the unusual task that it faces. No other human cell has to deliver a payload, in this case a package of DNA, the genetic material, into a foreign body. What is more, it is a foreign body that is out to kill it, since in the hostile environment of the woman's body the sperm is seen as a foreign invader that must be destroyed. Those sperm that survive this hurdle have to find and recognize one cell in this body, which involves swimming the equivalent of forty or fifty miles up the reproductive tract. They then compete to penetrate the membranes of the egg. Of the several hundred million sperm released in a single ejaculate, only a few dozen finally reach the egg. The larger the healthy army of sperm, the more likely it is that fertilization can occur. If sperm counts fall too low, and if their quality is impaired, the odds against successful fertilization begin to stack up.

Although sperm counts were falling, there was less evidence of a decline in male fertility. Several surveys showed that the number of couples seeking treatment for infertility had increased in the last twenty years[22] But these were less conclusive and any number of social or lifestyle factors could account for this increase. Unless large numbers of men reached the critical 20 million figure, there would be little impact on fertility.

But although the data on infertility was inconclusive, a decline in sperm count could only have one effect: it could only impair fertility. For an individual who already has a low sperm count, or poor sperm quality for some reason, an effect like this could tip the balance between being fertile and infertile. James McGowan's story is typical of the emotional distress that infertility can cause. In his case infertility has taken a huge toll on his health and personal life.[23]

Coping with Infertility

"I'd umpteen tests and it had taken months and months and eventually we got an interview with the doctor. In the course of about two minutes he destroyed my life. He said, 'Unfortunately your sperm count is so low that you'll never be able to have a child with your own sperm. You can adopt, but you'll never be able to have your own child.' I went away and sat in the car park for an hour, quite unable to move. I was just eighteen. Being told that I would never have a family was unbelievable. You never hear of anybody that didn't have kids, and so why me? I just sat there. I was devastated."

That was ten years ago and James's relationship with his girlfriend did not last. "It put a lot of pressure on us, because we both wanted kids. We started to argue about the tiniest thing. But we never spoke about what happened at the hospital. The whole thing was just taboo."

Eventually, he found a new partner who was sympathetic to his problem. They were both optimistic that they could seek treatment and soon be married. "We tried everything," James recalls. "When we first went to the hospital we were told: vitamin B, no hot baths, no tight underwear. It made no difference. But there was nothing else. They said they had made no progress in the treatment of infertile men. The only thing they could do was artificial donor insemination, which we also tried for two years. That meant going to hospital every month for more tests. We tried to find out when Hazel was going to ovulate and they gave us a kit that we had to inject into Hazel. This didn't seem to work either."

"There were so many tests," says Hazel McGowan. "The hospital were telling us when to have sex and it took a lot of the pleasure out of it. After about two years of trying, suddenly I became really depressed. I couldn't even go to work. It's like all of a sudden you find yourself at the bottom of a black hole and nothing matters anymore. You just want to make the world go away and you can't understand how you feel like this. I know I was depressed because we tried for so long to have a child. You have the feeling that something is missing, not just from your life, but from your body. Women have a maternal instinct and I felt if they could just take my maternal instinct away I'd be all right. I wouldn't feel like this anymore. It's hard to explain the feeling. There's something not there. You go out and see wee babies in prams and feel if I could just have a wee cuddle of that baby for five minutes it would make this feeling go away. But your arms are empty."

"I do feel I've let my wife down," says James, "because we are not really a whole family. We don't do a lot of things because there's no children. When we are walking down the High Street, we've spent years running past Tiny Togs shop window because we didn't want to look. . . . It hurts the most when we have had a niece or a nephew to stay and then they return home. You wake up, you're lying there, and it's really quiet. You're expecting to hear a child cry but there is nothing: no sounds of moving or snoring. Even the light doesn't move because nobody's jumping up and down upstairs. The house is silent and still. That's when it's really sore. . . ."

Dr. Stewart Irvine, consultant gynecologist at the Royal Infirmary of Edinburgh, explains that infertility is a very major problem. "It is very common, affecting about one in six couples in Britain," he explains. "Although traditionally it has been viewed as a problem that women suffer from, in about 40 percent of the couples that we see, it is the male partner who's the principal contributor to the couple's infertility."

"There are two basic problems with male infertility," explains Dr. Richard Sharpe, from the MRC Reproductive Biology Unit, Edinburgh. "The first is that there are very few treatments to offer to men who are affected. The second is that for men who are infertile it is perceived as a threat to their masculinity and virility. In fact there is no straightforward relationship between the two. You can be extremely macho and yet be completely infertile and vice versa. What determines virility and masculinity is a hormone called testosterone, the male sex hormone, which is responsible for all the differences between male and female, for larger size, greater muscle mass, deeper voices, and so on. When you survey infertile men, hardly any have a deficiency in testosterone. They've got all the male hormone they need to make them normally masculine. It's something else that has gone wrong with their process of sperm production that makes them infertile."

Tests are done on the sperm to establish not just the numbers but also the motility, or movement, and the percentage of normally shaped sperm. This can be done by measuring how easily sperm swim through a column of a chemical which is similar to the cervical mucus in viscosity. The sample can then be checked for biochemical abnormalities; sometimes the sperm itself can generate biochemical changes that reduce its ability to fertilize. Finally, it is possible to work out the fertilizing ability of a sample. The sperm are then prepared and the partner's egg is fertilized in the laboratory before being implanted using standard *in vitro* fertilization techniques.

While Niels Skakkebaek, Elisabeth Carlsen and the team were preparing their study for publication, Skakkebaek was invited to coordinate a World Health Organization conference on fertility in Copenhagen. Over a hundred specialists from all over the world would be invited. This, they decided, would be a good forum to present their data on changes to male reproductive health. If there was any flaw in their analysis, this would be a good place to discuss it. Some of the guests were people Skakkebaek had known and respected all his scientific career.

The group of experts gathered at the World Health Organization regional offices in Copenhagen in early October 1991. Over the course of three days there were many speakers and Niels Skakkebaek himself was very busy chairing sessions and coordinating events. He had to organize the whole meeting. His team were all assigned tasks to ensure the smooth running of the conference. In the evening there were social gatherings, dinners and functions to attend in the university. His talk, "Is human semen quality declining?" was scheduled for 9:30 on the Tuesday morning. It had already attracted a great deal of interest. The auditorium was packed. Rumors had spread that Niels Skakkebaek had some interesting new results. His reputation was such that everyone wanted to hear them.

"Our evidence suggests there has been a genuine decline in male reproductive health over the past fifty years," he began. "We systematically traced all the publications concerning semen quality among healthy men since the 1930s by means of a computerized search in Medline [an on-line medical database]. . . ." He was, as usual, gently spoken and intense, with typical scientific caution, the master of understatement. It was a lifetime's observation and study, summarized in just half an hour, in his slightly hesitant Danish accent. The seriousness of the tone and the impressive detail of the research only served to heighten the significance of the extraordinary results. "When Niels starts talking about his work, the excitement is infectious," explained a colleague. "He's so absorbed in the science, he has a way of penetrating to the essence. Everyone thinks it is the most exciting topic in the world."

He had the rapt attention of the audience as he explained their results on the slides in the darkened hall, the thin white line of the graph ever downward. "We have no reason to believe that our findings are due to selection bias or changes in methodology used for sperm counting. Thus the data provide evidence for a genuine decline in semen quality during the last fifty years. . . ."

His talk had extraordinary impact. Many in the auditorium were inspired to carry out their own studies. "He really woke people up," recalls Professor Frank Comhaire, from the University Hospital in Ghent, Belgium. "There were already bits and pieces of evidence suggesting that something was affecting sperm production. But such results as these, coming from Skakkebaek in particular, because he is known as a very, very serious man, created a great deal of interest. There were a lot of questions."[24]

Niels Skakkebaek, who had half expected the audience of experts to turn down his ideas as something which would be quite frankly impossible, found they were not as skeptical as he had feared. There was a slight time lag and then suddenly his study made headlines across the world from Tokyo to Los Angeles. "Today's man is half the man his grandfather was," announced Professor Louis Guillette to the delight of the press, as he gave evidence to Congress. In Paris, London, Edinburgh, New York, Helsinki and Brussels, scientific teams rushed to design their own studies to find out if manhood in their countries was still intact. What everyone wanted to know was, were sperm counts falling, and if they were, more importantly, what was causing it?

"Niels coped with it all very well," recalls Dr. Ewa Rajpert-De Meyts. "There were journalists even phoning his home in the middle of the night, forgetting the time zones. He's a very calm person, very down to earth, not easily fazed by critics, and so on. I even felt there might be a little hint of jealousy from other scientists in the field. He gained such media interest. He was even recognized on the plane. I think Niels sees it as your duty as a scientist, whether right or wrong, to state what you believe and I credit him with changing the style of thinking in this field. His studies have had tremendous impact. He has created a new wave of interest in the science of reproduction."

Later that year, Niels Skakkebaek was awarded the prestigious Novo Nordisk prize, the highest honor in Denmark for science, for his contribution to our understanding of testicular cancer. The man who had begun his career as an unknown, whose papers were rejected by leading journals for not fitting the scientific thinking of the day, was now being cheered and applauded by some of the most distinguished in Denmark.

Yet there were those who thought he had gone too far with his latest study. Behind the scenes, moves were already in progress which would attempt to undermine him and challenge his worldwide reputation.

Some time later, one evening when Niels was relaxing at home, a stranger called at his house.

"I saw your picture in the papers and I thought I would come and say hello," the stranger said. Skakkebaek realized the man was holding a bottle of wine.

"Come along in," he said. "Have a glass. Let us sit." They went into the kitchen and poured the drinks.

"Do you recognize me?" asked the stranger. "I remember you very well."

"To be honest, no, but I thought you would introduce yourself after you sat down."

"I was the first patient you treated, with carcinoma *in situ* . . ."

Suddenly Skakkebaek remembered, across the stretch of thirty years. The intensity of that small enclosed room where he, as a terrified young doctor, had to break the distressing news of cancer for the very first time. This was, of course, the patient whose cells had led to his discoveries of testicular cancer and set in train the ideas that influenced his whole career.

"I saw your picture in the papers. I came to find you. I wanted to thank you for helping to save my life."

CHAPTER TWO

The Paradox

Edinburgh, Scotland, 1992

Shortly after the World Health Organization conference in the autumn of 1991, Niels Skakkebaek was visited by one of Britain's leading experts on male fertility, Dr. Richard Sharpe. Sharpe had worked for many years at the MRC Reproductive Biology Unit in Edinburgh. As leader of a research team, he already had a considerable reputation for his contribution to the field.

Skakkebaek's wife had arranged a cocktail party at their home for the scientists attending the meeting that day. Sharpe and Skakkebaek, who had worked together before, soon found themselves discussing the events of the World Health Organization conference. "I had always admired Richard's work," says Skakkebaek, "and I was interested in his opinion of the paper we were writing about adverse changes in reproduction."

Against the warm ambience of the cocktail party, the strict discipline of their science held them totally. "Niels was very convinced this was a real drop in sperm counts. He's not a person who is given to exaggeration. He's very cautious," recalls Sharpe. "So when someone like that explains this to you, with quite a degree of personal conviction, even if hedged around with caution, I took some notice." Neither of them could possibly have imagined that their chance conversation would lead to a collaboration and a set of ideas that would attract worldwide interest.

At the earliest opportunity, they pored over the data, the evidence and its implications. "What this suggested to us," thought

Sharpe, "was that something has happened in the process of sperm production, this process that we know very little about, and this effect is not a small effect but a big effect and it appears to be a general effect. Something is going on, something big, and we have no idea what is causing it." Everything around them was unchanged, as it had been for centuries, the intoxicating charm of old Copenhagen. The uneven houses with their brightly colored façades, the old sailing boats in the harbor, the salty sea air. It seemed hard to believe that they were discussing such a momentous hidden change that might affect half the population. There were other concerns, too, nagging insistent questions that they could not yet answer. If the decline was real, had it stopped? Could sperm counts go down further? And if something could be having such a dramatic effect on sperm production, what other effects could it be having on us?

"I came away from Copenhagen thinking this was something I'd been half dreading," Sharpe was to recall later. "We'd been aware that our understanding of the process of sperm production was so limited. We have very little idea of what can make it go wrong. This puts us as scientists in a very perilous position." When he returned to the labs in Edinburgh, he informed the director of the Reproductive Biology Unit and others involved in male reproduction. These dramatic results were hardly the sort of news to remain secret for long. The puzzle was how to start investigating the phenomenon. In the fifty years that sperm counts appeared to have declined, our lives have changed immeasurably. Which factor, in all these changes, could be responsible?

His team were expert on the process of sperm production in the adult: spermatogenesis. The first question was whether anything could be damaging the developing sperm. "Spermatogenesis is a beautifully organized but very complex process," explains Sharpe, "which involves preparing and packaging the genetic material (the chromosomes) from the father, that will form the start of the next generation. It is a process, fashioned by evolution to ensure only the best quality copies of the genes are passed on. As a result the process is uniquely sensitive to anything that might harm the developing sperm."

Sperm start as a "germ cell," a round cell with a nucleus containing the genetic material or DNA a man has inherited from his parents: 23 chromosomes from his mother and 23 from his father. Firstly, through a series of cell divisions, these germ cells must lose half their chromosomes, so that when the sperm fuses with the egg it forms a cell with the correct amount of genetic material. In addition, as they mature, their shape is transformed. The remaining chromosomes are tightly packaged into a very compact oval head. A tail develops, so that they can swim. Everything superfluous, like cell proteins and other material, is shed so that they are as light as possible. This is to enable the sperm to travel the long distance to the egg with their important payload of genes. During this time, they also multiply in number and move from the edge of the seminiferous tubules in which they made toward the central lumen. It takes about ten weeks to transform the germ cells into the cells that can become the start of the next generation.

This process is particularly sensitive to a whole variety of environmental factors. To ensure only good quality copies of DNA are passed on, a whole number of checks and balances are in place, acting as quality control, constantly surveying the DNA. If something is found to be wrong, that cell may be pushed into programmed cell death. Radiation, for instance, is known to target dividing cells. Somehow, it triggers changes which make those cells abnormal. In turn, this is recognized by the body and the cell is destroyed. Programmed cell death is a means of removing damaged tissue without inflammation. After irradiation, under the microscope, the dividing germ cells which would normally develop into sperm have simply vanished.

Heat also causes damage. Tight underwear, cycling or driving for long periods, sedentary occupations, hot baths, tight jeans: all these have been blamed, since they all can raise the temperature of the testes. This damages the developing sperm another way. The cells in the testes are conditioned to operate at a lower temperature outside the body. It is thought that if the temperature is raised, this in turn will raise the metabolic activity of the cells, increasing the demand for oxygen and nutrients. If this demand outstrips supply, the forming sperm are damaged through lack of oxygen and nutrients. Once again, if damaged, the body will dispose of them.

Apart from heat and radiation, certain chemicals, too, were known to affect the cells of the testes. Occupational exposure was known to be a risk. But none of these factors appeared to account for all the adverse changes they had seen. "It was not sufficient to explain anything," recalls Richard Sharpe. "We sat down and talked about this, saying what is there that could have changed in our environment over this time period that is actually lowering sperm counts in the adult, day in day out, week in week out, and doing this on a population scale. We just couldn't identify anything. It is difficult to believe that heat, radiation or the variety of chemicals that are out there could actually have had such widespread effects on a large proportion of men. I think it was at that point that we began to question whether the decline in sperm counts could be occurring in some other way."

There was, after all, not just the possible fall in sperm counts to consider, but also the other adverse changes. "You could never explain the rise in testicular cancer by tight underpants, or something happening to the adult," explains Skakkebaek. "Testicular cancer is almost certainly derived from changes to the fetus, during development." They began to wonder whether all these changes to reproduction could in some way be related. The developmental abnormalities in baby boys and the data on testicular cancer suggested something was happening early in life. Was it possible, they wondered, that events early in life could also affect sperm count?

By chance, Sharpe, who was editor of the *International Journal of Andrology,* was sent an article to review early in 1992. It had a characteristically lengthy title, full of scientific promise—"Induction of increased testis growth and sperm production in adult rats by neonatal administration of the goitrogen propylthiouracil: The critical period."[1] To anyone else it might have seemed indecipherable jargon of little significance, a mere scientific curiosity adding to the piles of scientific data that accumulate each year. But Richard Sharpe, perusing the article in the routine clatter of the lab, could see its significance immediately.

In essence it outlined a complex piece of research which showed that it was possible to alter sperm counts in adult animals by altering hormone activity during development. "The researchers had suppressed the activity of the thyroid gland in newborn rats for the first two weeks of their life," Sharpe explains. "When you suppress activity of the thyroid gland, you slow down growth and development so the animals actually grow up a little bit smaller than normal. But remarkably, despite this, in adult life these animals had larger testes and higher sperm output. In fact, by suppressing the thyroid, they could get the testes almost twice the size. When they looked down the microscope, they found the process of sperm production was completely normal and they concluded the most likely change was in the number of *Sertoli cells*."[2]

So if, he reasoned, sperm output could be *increased* by increasing the number of Sertoli cells formed, could the reverse also be true? Would sperm counts in the adult be *decreased* if the number of Sertoli cells formed during development was lowered for some reason?

"Sertoli cells," he had always told his students, "are probably the most important cells in the testes. They are the cells that control the whole process of sperm production. Each Sertoli cell has 'germ cells,' the precursors of sperm, embedded in it. There's a limit to how many germ cells you can embed in a Sertoli cell. So, in effect, the number of Sertoli cells determines how many sperm you can produce. The greater the number of Sertoli cells, the greater the number of sperm produced. All the variations in testicular size that we know that occur between animals are due to differences in numbers of Sertoli cells."[3]

Sertoli cells act like a nursery cell, supporting the developing sperm. Under the microscope they are laid out rather like a Christmas tree. The germ cells, the precursors of sperm, are jammed in between the feathery branches of the Sertoli cell. It has a sort of trunk in which the nucleus is embedded. He had studied them so often, but had never thought to investigate the process of development of the Sertoli cells. "And it was this realization that suddenly seemed absolutely critical. Up till that moment I had thought the *development* of Sertoli cells was unimportant and of course I was completely and absolutely wrong," he says with a smile. "Now I tell everybody in the lab that if I say something's unimportant then take a note, because it'll probably turn out to be the exact opposite."

He rearranged his commitments for the afternoon, and went to a specialist medical library, the Erskine Medical Library, in Edinburgh to check out everything that was known about Sertoli cells during development. He soon uncovered two remarkable aspects to the process. Firstly, there are critical periods when Sertoli cells are formed. These periods occur in the womb and during early childhood. At some point in childhood or perhaps early puberty, the numbers become fixed and no more multiplication is possible. Because the number of Sertoli cells determines the number of sperm that can be produced, an individual's ceiling of sperm production as an adult has already been largely determined before puberty.

So, he wondered, could some unknown agent be fixing Sertoli cell numbers early and reducing sperm counts? Over the course of the next few days he continued a detailed search through the literature. Gradually, he began to put together a picture which astonished him.

"Secondly, it became clear the Sertoli cells are really in the driving seat during the entire development of the male reproductive tract in the womb. By their secretions, they control all the other processes of development of the male. Take, for instance, the descent of the testes from the kidneys: part of this process is signaled and controlled by the Sertoli cells. The Sertoli cells also act on the germ cells, the cells that make the sperm later on in life, to suppress them so they don't start developing into sperm until puberty. If something interferes with this process of suppression it's thought that this may lead to testicular cancer later in life. Sertoli cells also control other cells, known as 'Leydig cells,' which make testosterone. This makes the boy masculine and is involved not only in the descent of the testes, but also in the development of the reproductive tract and the external genitalia.

"Suddenly we had this remarkable bringing together of all these facets of the story, so that testicular cancer, testicular non descent, urethral abnormalities and, via Sertoli cell numbers, sperm counts in adult life, all appeared to be related through the Sertoli cells. If something was affecting the function and number of Sertoli cells, that could explain all the abnormalities that we were seeing." But what?

He went back and reread all the articles. There was one particular study that held his attention. It was outlining the effects of estrogen, the female hormone, on the development of Sertoli cells. The female hormone, this suggested, may play a vital role in the development of the *male*.[4]

Hormones are incredibly potent chemical messengers in the body. Present in tiny amounts, parts per billion, they can nonetheless communicate with the powerhouse of the cell, the genes, in the nucleus. The genes contain the set of instructions, inherited from our parents, that makes us what we are. Hormones help to determine how these genes express themselves; they can mask genes, and turn them on or off. In this way they play a fundamental role in expressing what we are, and shaping the countless processes essential for life.

Estrogen in the female is secreted by the ovary and regulates development, growth and reproduction. At puberty, estrogens are responsible for the secondary sex characteristics in women, the enlargement of the breast, deposition of fat on the hips and thighs, and the appearance of pubic hair. They also regulate the monthly cycle, whereby the egg develops and is released from the ovary (ovulation) and the womb is prepared for implantation. During pregnancy, estrogen levels rise sharply. During menopause, the decline in estrogen levels can lead to wasting of the genital tract and contribute to bone loss and the development of osteoporosis.

Sharpe studied the extraordinary intricate process by which the fetus develops into a male at around six weeks of age. It involves a series of carefully orchestrated hormonal cues where timing was critical, involving male and female hormones. There was the follicle-stimulating hormone released by the pituitary, which in turn drives estrogen production by the Sertoli cells and the formation of Müllerian-inhibiting substance, a hormone which makes the female ducts regress and disappear. At the same time, testosterone, the male hormone, is released by the developing testes and drives masculinization of the reproductive tract. As he pored over the articles unraveling the complex relationship between hormonal secretions and key events in the development of the male sex, he suddenly saw that there was one common factor that could disrupt the whole process: excessive exposure to the female hormone estrogen during development.

Estrogens, by suppressing other hormones could reduce the multiplication of Sertoli cells and fix their numbers early, at low levels. This would in turn reduce sperm counts in adult life. In addition, interfering with the secretions of the Sertoli cells could also, depending on the timing, affect descent of the testes and the development of the urethra—a critical factor in hypospadias—and set in train the events that lead to testicular cancer. Estrogens, it appeared, could throw the entire delicate system out of balance and mess up development permanently.

As he thought of this, he wondered if he'd made a mistake. After all, every fetus is of course exposed to its mother's natural estrogens in the womb, yet this doesn't derail development. What is more, during pregnancy, estrogen levels are at an all-time high. His own conclusion seemed so paradoxical. But as he checked all the data, he realized that the fetus may not be exposed to all of the mother's natural estrogens after all. Because hormones are so potent, we have evolved a way of transporting them in which they are inactive, until they reach the target cells. Our natural estrogens bind to a protein in the mother's blood, sex hormone binding globulin (SHBG). This and other mechanisms are thought to protect the fetus from the mother's sky-high hormone levels. But, it occurred to him, what if something else, *some unknown agent,* was acting like an estrogen in the womb and crossing the placental barrier: it could disrupt the entire sex development of the male fetus.

This was quite a moment. "Quite frankly, I was astonished to have taken all these disparate pieces of information, which were already there embedded in the literature, and to come up with a theory that seemed to explain all these adverse changes we were seeing. It was a time of real discovery," he recalls. "Obviously there's tremendous elation, as all these pieces of the jigsaw suddenly fall into place. It is the old eureka thing, you know: you want to jump up and down. Then you begin to worry about the consequences and you think, oh my god, what is it that I've stumbled upon and how can I actually firm it up. It's a great thing to have ideas, but then you've got to check out whether they are real. . . ."

Was this just a tenuous thread, a mere hypothesis in which he should place no trust? While it was shocking and exciting to see so clearly the pathway whereby estrogen exposure could harm human reproduction set out in plain print before his eyes, he still wondered if he could be wrong. Was he about to make an almighty fool of himself? Or was he the first to ask the right questions? He went back to check whether there wasn't some important information he had missed. But he could see no flaw.

If this was right, there had to be something out there, all around us in the environment which could act like an estrogen and to which we had unwittingly been exposed. It took millions of years for humans to develop the unique adaptations that we have today. Millions of years of evolutionary heritage, which perhaps in just fifty years we may have altered. But what was the cause, and could he prove the connection?

"When my feet came on the ground, I began to think of all the things I didn't know, all the flaws that may be there in my arguments, and that meant finding a way of putting these ideas to the test," he recalls. Over the course of the next few weeks, he discussed his ideas with colleagues and, of course, with Niels Skakkebaek. They soon realized that there had been an accidental human study of synthetic estrogen exposure. Up to six million babies were exposed to the synthetic estrogen DES between the 1950s and 1980s, when it was given in the belief that it prevented miscarriage. Later it was found that, for many of the offspring who had been exposed to DES in the womb, it was to be a tragedy that would devastate their lives. A great many studies were carried out on those exposed to the synthetic estrogen DES, studies which they realized would hold the key to their ideas.

The Legacy of DES

The case of DES is seen by some as a distressing, albeit unintentional, human experiment on six million pregnant women and their offspring. For them, the story has become a symbol of the harmful consequences of inadequate drug testing and the willingness to experiment on women. For others, the tragedy is merely part of the learning process of science, reflecting only our ignorance and willingness to take risks in order to learn and progress. However it is viewed, the story of the drug DES has its origins in the early part of this century.

In the 1920s, the study of hormones, known as endocrinology, was in its infancy. It was recognized that certain organs in the body, such as the ovaries and the testes, had powerful secretions that appeared to control many aspects of health: development, growth and sexuality. But no one knew what they were, or how they worked. No one could even isolate the active substances which possessed these fantastic powers and they remained elusive for many years.

Meanwhile, ecstatic reports from commercial laboratories promoted miracle cures obtained with extracts of animal glands. The powerful secretions from these glands were sold as cures for aging, sterility, frigidity, problem pregnancies and an infinite variety of ills. In short we could defy nature with these magical potions. Extracts of animal ovarian tissue, perhaps mixed with a little spleen or pancreatic tissue, were touted as an "elixir of youth." Secretions of monkey glands would guarantee male potency. For the serious scientists trying to place the study of hormones on a firm scientific footing, the wildly overoptimistic claims of the charlatans hindered their progress. "The results of incomplete experiments and fantastic hypotheses are thrown together to form a glittering and ever-changing kaleidoscopic picture," complained one scientist, Professor Robert Frank, in 1922. "Shotgun mixtures are being circulated. What is to be the end of this seemingly uncontrolled wave of mysticism, hysteria, commercialism and credulousness?"[5]

The end came gradually as our scientific knowledge progressed. The female hormone, known as the "hormone of estrus" in the 1920s,[6] was shown to be produced by all female mammals. It was produced by the ovary and regulated the maturation of the egg in the ovary. But no one could isolate it. "This was, at least in part, because of its incredible potency," recalls Professor Harry Jellinck, who was involved in some early studies of estrogens. "It is present in the ovary in very tiny amounts. The scientists would have needed some ten tons of ovaries to obtain a few milligrams of estrogen. It was a tremendous task to isolate the substance. Chemists were pushed to the limits of their ability. It was the most potent of any substance known at the time."[7]

Eventually, however, the active ingredient from the ovarian secretions was isolated by E. A. Doisy at Missouri University, who later won a Nobel prize for his work. It became known as estrogen because it produced the state of "estrus," where the female animal is in heat. This occurs after ovulation, when the egg is released from the ovary and the womb is ready for implantation. Estrus is derived from the Greek *oistros*, meaning "frenzy."

Once estrogen was identified, analytical chemists wanted to know how it worked. Why was it that a molecule of estrogen had these extraordinary properties that enabled it to regulate reproduction and growth? A British team, headed by Charles Dodds, at the Courtauld Institute of Biochemistry in London, began to study substances that had similar structures to estrogen to see if they too had similar properties. It was research which has been widely admired and which made the team famous, although it was unwittingly to lead eventually to the DES tragedy.

Professor Jellinck, who later worked with Dodds, recalls the story: "They took compounds off the shelf and analyzed them to find out what they could do. In the 1930s they became interested in a family of chemicals called the stilbenes, which showed some estrogenic activity. They also wanted to know whether any simpler compounds were active and were very excited to find that one of these, propenylphenol, appeared incredibly potent as an estrogen. In fact it was a fluke. They found that different batches had different activities which made no sense at all." They realized their sample was contaminated and began a series of experiments to work out the nature of the impurity. This trail led them to isolate the first synthetic estrogen: a substance called *diethylstilbestrol*, or DES.

Remarkably, it was as active as the most potent natural estrogen, estradiol. The team announced their findings in a letter to the prestigious journal *Nature* and the medical community immediately caught on to the excitement.[8] In scientific circles this was an important discovery. Dodds was acclaimed by the British establishment and later received a knighthood for his efforts.

"He had a fantastic reputation," recalls Jellinck, who joined his team later as one of his few graduate students. "He eventually became President of the Royal Society and was very well regarded. His list of honorary degrees was a mile long. In fact, he was really quite a shy person, very well protected by his secretary, and I don't think particularly interested in the notoriety."[9]

Dodds himself was fascinated by the scientific implications: what was the mechanism that enabled DES to mimic estrogen? But others were more interested in the commercial implications. As the news soon spread beyond scientific circles, it was realized that a synthetic estrogen could now be used in treatment. What is more, since Dodds, in true British style, failed to patent his discovery, any manufacturer could try to make money from his invention. Suddenly, the number of symptoms that might be amenable to estrogen therapy seemed enormous. Less than two years after Dodds's letter to *Nature*, America's drug regulatory body, the Food and Drug Administration (FDA) had received over ten requests from pharmaceutical companies to market DES.[10]

According to Anita Direcks and Ellen't Hoen in their report on DES, the FDA were instrumental in the launching of the drug. At first, FDA officials were not satisfied with the data from the manufacturing companies. Individual requests from manufacturers were withdrawn and a joint application was set up. About eighteen months later, the FDA was informed of a large study involving over 5,000 successful treatments with DES. This time, in a matter of weeks, the FDA officials were satisfied with the data, and approval for the commercial use of DES went ahead. Curiously, an FDA official involved in the approval of DES later became president of one of the first chemical companies to manufacture DES.[11]

FDA approval came in spite of early warnings of possible carcinogenic effects of estrogen treatment. During the 1930s, a French chemist, Lacassagne, was able to demonstrate that estrogen exposure could induce breast cancer in male mice. Perhaps because these findings were published in French, they did not attract much attention.[12] However, there were other clues that the sex hormones may have some relationship with cancer. It was already known that if women had their ovaries removed, and therefore were not exposed to any estrogen, they rarely developed breast cancer. However, these early warning signs do not appear to have influenced the FDA decision. With hindsight, it would appear that there was very little research carried out on animals. Women themselves were to be the guinea pigs.[13] It was to be a disaster.

Over the course of the next decade a great many uses of the drug were found. First, it was given to alleviate menopausal symptoms, replacing the loss of estrogen at menopause. Then it was found that a large dose of estrogen would suppress lactation and that it was also effective as a contraceptive: "the morning-after pill." It was used in the treatment of breast cancer and prostate cancer. Fantastic claims began to spread about this wonder drug. Because estrogens can promote growth, it was even used in some countries as a hair growth tonic and in "sex pills" to encourage potency and arousal. In agriculture too, its effects appeared miraculous. In the early 1950s it was shown that when given as an implant or a feed additive it could increase the daily weight gain of the animals, increase food consumption and increase the efficiency with which food was used by the animal. Studies showing that DES-treated meat was less tender, with larger and more coarse muscle fibers, were ignored. DES could fatten animals fast and was used in farming from the early 1950s to the mid-1970s.[14] But the most tragic use of DES was in pregnant women.

During the 1940s, a husband and wife team, George and Olive Smith, were studying hormones in pregnancy and found that, when complications arise, there can be a drop in estrogen levels, triggering a drop in progesterone as well. Their research suggested that DES might be able to restore normal hormone balance in pregnancy and they proceeded to check this out. Sure enough, by 1948, they reported a study of 632 pregnant women which seemed to show that DES did indeed have a beneficial role in high-risk pregnancies, helping to prevent complications such as abortions, premature delivery, pre-eclampsia and intrauterine death. They believed DES stimulated estrogen and progesterone production, making a healthier environment for the fetus. Despite the lack of adequate controls in their study, it became highly influential.[15,16] They began to promote the use of the drug for the treatment of habitual or threatened abortion. According to some physicians, the doses they recommended were large.[17] "Five milligrams daily by mouth is started during the 6th or 7th week . . . this is increased by 5 milligrams at two-week intervals to the 15th week, when 25 milligrams are being taken daily. Thereafter the daily dose is increased weekly by 5 milligrams."[18] Following their regimen, sometimes doses as high as 12 grams of DES were taken in pregnancy.[19]

No proud mother-to-be looking at the advertisements for the drug in the late 1950s could have guessed what lay ahead. In 1957, DES was even advertised in the medical press. A wide-eyed and smiling baby, a glowing picture of health, was featured and the advertisement proclaimed: "Yes, DESplex to prevent abortion, miscarriage and premature labour, recommended for routine prophylaxis in *all* pregnancies." The advertisement promised: "96% live deliveries and bigger and stronger babies too and no gastric or other side-effects with DESplex." The understandable impression for many mothers was that you were not doing the best for your baby if you did not take the substance. It is estimated that between 1951 and 1953 as many as 10 percent of all American babies were exposed to DES.

The benefits of the drug during pregnancy were first questioned in 1953 when a massive study on nearly 2,000 women in Chicago showed women taking DES were *no less likely to miscarry*. William Dieckmann from the University of Chicago wrote: "To ameliorate some of the common hazards of childbirth for mothers and babies is certainly a worthy goal. That all this can be accomplished by the daily consumption of a few tablets is indeed enticing." However, he considered the Smiths' studies flawed in a number of respects, as indeed was the theory that a dose of estrogen would in some way regulate hormone levels in pregnancy. His detailed and carefully controlled study showed no benefits in taking DES.[20] But sadly, despite his findings, the drug continued to be prescribed in the West. Eventually up to six million mothers and babies were exposed.

The bombshell was to come fourteen years later. In 1970 Professor Arthur Herbst at the Vincent Memorial Hospital in Boston was astonished to find that within a three-year period he had seen seven teenage girls in the Boston area with an exceptionally rare type of vaginal cancer: adenocarcinoma, or clear-cell cancer, so named because of the appearance of the cells under the microscope. No such cancer had been reported at the hospital before. In fact, previously the condition was so extremely rare that only three cases had been reported in the entire international literature. The teenage girls had been referred with abnormal bleeding. Their diagnosis couldn't be more traumatic. To save their lives, surgery was necessary. This involved a radical hysterectomy, removing the vagina and womb so they could not have children, and later rebuilding the vagina. Several of the girls had more advanced cancers. Radiation, chemotherapy and more radical surgery were carried out. In some cases the prolonged bleeding caused by the tumor had been mistaken by their doctors for period problems and this had delayed referral and diagnosis. One of the girls had such an extensive tumor she could not be treated surgically and died within a few months.[21]

Herbst and his team were mystified about this apparent cluster. His colleague Dr. Howard Ulfeder later described what happened: "We were on the watch for an explanation, but none suggested itself until the mother of one of my patients reported as additional past history that she had been prescribed DES during pregnancy to minimize the chance of loss of the fetus. . . . Somewhat to my surprise, I found on questioning that several other mothers of the group had also taken DES during pregnancy. . . ."[22] Soon it became horribly clear that the seven young women all had one thing in common: their mothers had taken DES during their pregnancy.

In April 1971, Arthur Herbst reported these disturbing findings in the *New England Journal of Medicine*.[23] Over the course of the next few years, it was realized that DES had many effects in the female offspring. The uterus, vagina and fallopian tubes could be misshapen and deformed so that pregnancy was not possible. These abnormalities can include underdevelopment of the cervix or uterus, or misshapen organs, cervical hoods and collars, or a T-shaped uterus. Glandular tissue normally found inside the cervix would be found on the outside. Not surprisingly many of the DES daughters had difficulty reproducing, with spontaneous abortions, ectopic pregnancies, stillbirths and premature births. There was increased incidence of cervical and vaginal dysplasia in the young women, a precancerous condition marked by disorderly cell changes. But worst of all were the higher rates of cancer in the DES daughters. Even with treatment there has been a 20 percent rate of recurrence—sometimes not in the pelvic area but in the lungs or elsewhere. Most of the patients with persistent or recurrent disease have died.[24,25]

Charles Dodds, himself, who had first synthesized DES, was "quite upset" at what happened, recalls Jellinck. "It was rather sad, because by that time he was getting old and this was his key contribution to science, for which he had been so celebrated and admired as a younger man." As the news unfolded of serious adverse effects on women, Dodds became very concerned. He died soon after this.

For Richard Sharpe looking up the studies of DES, it was very clear that the drug had a dramatic effect on the female offspring. There was much less data on the male offspring. Sharpe tracked down the studies, because these would be critical in revealing whether estrogen exposure could harm male reproductive development.

In a review of all the data, Professor William Gill from the University of Chicago concluded: "DES exposure of human males *in utero has* resulted in an increased incidence of abnormalities of the male reproductive tract. . . ." A range of distressing abnormalities were described, each one carefully measured and couched in the meticulous unemotional language of science. These included increased incidence of cysts, abnormally small or undescended testes, testicular cancer, reduced sperm counts, hypospadias, and microphallus. Definitions of this last condition varied. In one study, where the average normal penis size was taken as 12 centimeters, a mild microphallus of 7 centimeters or less was found in over 30 percent of DES-exposed males; there were also some severe cases where penile length did not exceed 4 centimeters. There were even incidental reports of male pseudohermaphroditism. One infant who had received very high doses of DES was classified as hermaphrodite, with a microphallus, urethral abnormalities, and undescended testes. In all, although not all studies reported effects, there was a litany of distressing conditions, which for the individuals concerned could in some cases affect them for life.[26]

"Overall the studies showed fairly clearly increased incidence of testicular non descent and a greatly increased incidence of urethral abnormalities; they also tended to have lower sperm counts and poor semen quality in adult life. So it looked as though they were showing all the sorts of abnormalities we predicted," Sharpe concluded.

"The fit of the data seemed too good to be true, that was my immediate reaction," recalls Richard Sharpe. "Usually you have a situation where you don't have any human data and you obviously can't do experiments on humans. Yet here there was this very unfortunate episode where millions of women and their babies were exposed to a synthetic estrogen. Of course, it puts the pressure back on you, because you realize that what you are thinking does have potential significance."

Richard Sharpe and Niels Skakkebaek began to prepare a paper for publication in the *Lancet* outlining how exposure to estrogen could affect male reproduction. They were both acutely aware that, although all the facts fitted, it was still just a hypothesis. "What we wanted was to have our ideas exposed to scientific scrutiny in as wide a sense as possible," says Sharpe. "The more people that look into this, not only are they going to give us ideas, but they might spot possible flaws. Even if all our thinking is proved wrong, which would obviously be disappointing, for someone to prove it wrong, they are going to advance our understanding so much that we are all going to be ahead anyway." As it happened, the opportunity for scientific scrutiny came a little earlier than anticipated.

The Royal Society Summer Exhibition is one of the highlights of the science year. The grand white stucco building in the heart of Piccadilly, London, overlooking St. James's Park has been a center of support for scientific excellence for centuries. The organizers, with unflagging interest, would ensure that Britain's most brilliant and intriguing science was on display to stimulate debate, keep the profile of science high and draw attention to good scientific work. This was the science establishment keeping order. Anyone who thought he had stumbled upon something when in fact he hadn't would soon be called to account. By day the public would be allowed to view the exhibitions. At night there were special functions and celebrations. The country's leading scientists would be invited. The media would be present in force. Many famous scientists had presented their ideas here. So it was a considerable honor when Richard Sharpe and Niels Skakkebaek heard that they were invited to present their ideas to the Royal Society in June 1993.

The halls were filled with exhibits covering the latest exciting advances in molecular biology, nuclear physics, genetics and biology. Beneath the stern gaze of their scientific forebears, the brilliant could dazzle the up-and-coming, under the admiring gaze of the press, who were polishing their smiles while waiting for the impenetrable language of science to be decoded. This was a scene for London's elite: the brightest, the cleverest, the most inspired. Portraits of Sir Isaac Newton, Joseph Priestley, Charles Babbage, Lord Rutherford, Henry Cavendish, John Wallis, and many others hung in gilt frames on the walls. Four hundred years of scientific authority and tradition were reflected softly by the chandeliers and the mirrors, embodied in the glittering trappings of Carlton House Terrace. In this great hall of wisdom, they were half expecting someone to see a flaw in the hypothesis.

But they need not have worried. They found that the world of science was very receptive to their ideas. There were no immediately obvious pitfalls in their theories that they couldn't counter. No one could provide the comforting reassurance that they had got it wrong. Meanwhile the press metamorphosed the hypothesis into world catastrophe. Their careful, pedestrian, quiet scientific dedication was transformed into impending disaster, the death of the human race, the ultimate full stop. Estrogen, the female sex hormone, became the prime villain. The race was now on to identify the mystery estrogen to which we had all been exposed.

CHAPTER THREE

A Sea of Estrogens

North Carolina, 1993

Unknown to Richard Sharpe, at a research institute in the heart of the Deep South in America, another scientist, Professor John McLachlan, had been concerned about estrogen exposure for many years. He had reached an uncannily similar conclusion to the European scientists, but had arrived at this through a totally different set of experiments which had influenced his thinking for some time.

Now in his early fifties and at the height of his career as the distinguished scientific director of the National Institute for Environmental Health Studies, in North Carolina, the key steps in his thinking which led him to examine what he termed "environmental estrogens" had begun years earlier. It all started with a fascination for human development while a student at the George Washington University.

At that stage, the development of the embryo from one cell, to many cells, into the extraordinary complexity of a human fetus in the womb was, as he recalls, "this marvel of life, beyond scientific explanation." Until comparatively recently, scientists assumed that the womb was inviolate: a dark, secret place which couldn't even be visualized routinely by modern science, until the development of ultrasound in the late 1970s. In the 1930s, some X-ray images were taken, before scientists realized the harm this could do. These extraordinary black-and-white images, showing every last detail of the developing fetus curled up inside the mother, appeared to confirm the comforting perception built up over centuries: the mother, like some sort of Madonna, shielded the baby from all adversity. The mother's placenta, it was believed, provided the ultimate barrier, protecting the infant from any harm.

"The tragedy of the drug thalidomide overturned this comforting illusion of the inviolate womb, once and for all," recalls McLachlan. The drug was given to women in the 1950s and early 1960s during troublesome pregnancies. Many of the infants exposed to thalidomide in the womb were born with heart-breaking deformities, no arms and legs, or with tiny hands or feet growing where their limbs should be. The victims, babies and children, were displayed on the front pages of magazines for months, a horribly visible challenge to medical complacency. For a young John McLachlan, embarking on his scientific career, this tragedy led him to question how foreign chemicals could affect development, well before this became a fashionable field. His thesis adviser was Dr. Sergio Fabro, who, besides having three Ph.D.s to his name as well as being a world-class bridge player, had also done landmark work on thalidomide. "I specifically sought out Sergio. He had already made the intellectual leap—I'd like to think we made it together, but I think he made it first—that if chemicals caused defects in the womb, they must have done this through some biological process, and this could have consequences later."

His other tutor was Roy Hertz, who was very influential in the field of endocrinology, the study of hormones. He became a "long-term friend and mentor" and they had numerous discussions about how hormones work. What, for instance, makes an estrogen affect cell growth and have such profound biological effects? This inevitably led to discussions of DES, or diethylstilboestrol, the famous synthetic estrogen mimic synthesized by Dodds.[1] At this stage, in the late 1960s, there was absolutely no evidence that the drug, given during pregnancy when miscarriage threatened, had any adverse effects on humans. But McLachlan, aware of the worldwide use of the drug in the cattle industry for fattening livestock, was concerned. DES was a very potent synthetic estrogen. Was it wise, he wondered, to put 13 tons of it into the environment every year, before we know what effects it may have on development? He accepted a post at the National Institute of Environmental Health Sciences (NIEHS) at Research Triangle Park, North Carolina. His task: to study the effect of DES on development.

The building, secluded in its own grounds at the end of an empty road, was a vast edifice of red brick rising out of the dense forest. Conifers and pine woods stretched for miles, as far as the eye could see. The windows were dark glass, to keep out the glare of the southern summer. An obviously artificial lake had been built at the back. Prominent at the front of the building was an American flag. Undaunted, John McLachlan found he had his own laboratory and quickly assembled his own specialist team. The first person he hired was a cytologist, Retha Newbold, an expert in understanding disease pathology. The first doctoral fellow he appointed was Kenneth Korach. It was the start of a highly successful collaboration that would last for twenty-two years.

Soon after they were established, their research took on a new-found urgency. In 1971, Professor Arthur Herbst published details of a cluster of young women with rare vaginal cancer, all of whom had one thing in common: all were exposed to DES during their mother's pregnancy.[2] This became headline news all over America. It was the kind of news that McLachlan half expected, and half dreaded. "Suddenly there was a crisis medically in terms of the millions of women who had been exposed to DES." Their DES research was no longer just an intellectual curiosity. They suddenly found themselves in the front line of cutting edge medical research. "At first, it was not accepted that DES had caused the cancers. There was a heated debate," explains McLachlan. "People argued that these were problem pregnancies anyway, which the drug might have salvaged. Maybe they had chromosomal abnormalities, and this caused the vaginal cancers." Unlike thalidomide, which caused defects at birth, these defects were delayed until puberty or later. The vaginal cancers appeared to develop when the hormonal environment changed in adolescence. Such delayed effects of chemicals in the womb had not been reported before and the case against DES was far from clear-cut. This was also the first case of *transplacental carcinogenesis* that had been proposed, where chemical exposure in the womb caused cancer later. McLachlan recalls: "The

Herbst study revolutionized the way we thought about these chemicals."

In the laboratory, McLachlan looked again at the mice which had been exposed to DES during development, examining the reproductive organs in much more detail. To his amazement, he noticed alterations in the vagina, which looked remarkably like vaginal cancers in these mice. Vaginal cancer had never before been seen in a mouse. No one in the laboratory had sufficient expertise to definitely judge whether it was indeed vaginal cancer, so McLachlan sent the slides to a leading pathologist, Professor Bill Bullock, to provide an independent evaluation. If it was true, then they would be the first to have a mouse model of this human disease, which would be crucial in helping to understand and treat the distressing vaginal cancers in women. By now the scale of the problem of human exposure to DES was becoming apparent. Thousands of women were affected and every week there would be some new heart-breaking mother and daughter case in the papers.

McLachlan's position was under review. "I had a presentation coming up, and that had a lot to do with whether I'd be asked to stay." He needed an answer to the question of the vaginal cancers quickly.

"We put the slides in the trunk of the car. But the first slides were crushed to pieces. Someone put something on them. So then we rushed another set of slides round to the pathologist Bill Bullock. And he called me up to say, yes, he had evaluated the slides and he was willing to confirm that this was definitely vaginal cancer, which had never before been seen in a mouse."

McLachlan gave his presentation, explaining that his team had in effect developed a mouse model for a human disease. Suddenly he became the man of the moment. His position at the NIEHS was renewed and his work on DES became very prominent. "It was amazing," recalls McLachlan, "we just didn't at that time have the conceptual framework for thinking that chemical exposure in the womb could affect the fetus, not at birth, but much later, at puberty, when they come under other hormonal influences. It was an extraordinary biological outcome." Their discovery was to lend considerable weight to the idea that the drug DES had in some way caused the rare vaginal cancer that had been reported in young women.

As his team continued their studies, they found something even more unexpected in the male offspring. "The thing that really shocked and surprised us when we looked at the *male* offspring of these DES–treated pregnancies," says McLachlan, "was that they actually had both a male and female reproductive system existing side by side. *Essentially they were hermaphrodites.*"

This was a puzzle. Most vertebrates, including mice and humans, start fetal development in a bisexual form, with two coexisting male and female reproductive systems existing side by side. In humans, at around six weeks, the Müllerian duct starts to develop if it is to be a female and will lead eventually to the formation of the uterus and vagina. The male Wolffian duct regresses. However, if the fetus is male, the female Müllerian duct disappears and the Wolffian duct develops, eventually forming the testes, vas deferens and seminal vesicle.

"We finally figured out that this feminization process of the males occurred early in fetal life when they essentially exist in a bisexual form. DES froze the process, so that the fetus retains both the male and the female reproductive systems developing side by side. The male is feminized in such a way that it has both male and female systems intact as well as testes. There were no reports in the literature of such an effect at that time, so we were surprised to find these very dramatic effects on males." He was very concerned. So far, their observations in mice had been strangely predictive of what would be found in humans. They published their unusual findings in *Science* in 1975, reporting that 60 percent of the male offspring from pregnant mice treated with the estrogen DES were sterile.[3] This study soon became the definitive checklist for clinicians for the effects of DES on males.

But John McLachlan's team did not stop there. They continued to observe these male mice as they grew older and found other sinister changes. It was this work which prompted McLachlan to worry about whether there might be long-term and widespread effects from synthetic estrogen exposure. The evidence was beginning to suggest to him that events in fetal life could create a predisposition to disease later in adult life.

By chance, in 1987 a new member had joined that laboratory team, Risto Santti, Associate Professor of Anatomy from the University of Turku, Finland. He was a specialist in reproductive biology, especially human prostate diseases. At a conference in North Carolina, he had heard McLachlan talk about his research on DES and estrogen exposure. This was exactly the field which fascinated him and the two began to collaborate.

"The exact function of the prostate is unknown," explains Professor Santti. "It is located close to the neck of the bladder in men, surrounding the urethra which takes urine from the bladder. It secretes material which forms part of the semen and is thought to have some importance for fertility."

"When we looked at these male mice as they were getting older," recalls McLachlan, "around a year, or year and a half, which is old for a male mouse, and looked at their prostate glands under the microscope, we could see changes that suggested to us that their prostate might be in an early stage of what could eventually be a cancer."

Down the microscope they could see wedges of abnormal cells embedded in the surface or epithelial cells of the prostate tissue. At high magnification, it was possible to see cells with an enlarged nucleus, a characteristic of malignancy. There were also glandular changes, showing similarities to prostatic cancer in humans. "Several of the pathologists who looked at this with us said that these changes would be consistent later perhaps with much more severe disease," recalls McLachlan.

The mice had other symptoms too, all consistent with prostate disease: more inflammation, increased frequency of urination and decreased volume of urine. All of these changes are typical precursors of benign prostatic hyperplasia, in which the glands become enlarged and change shape in such a way as to obstruct the urethra. Since this restricts the flow of urine, it can lead to the uncomfortable symptoms and the difficulty in urinating. What is more, *all* the male mice showed these changes: it was a widespread, although subtle, effect.

These findings were worrying. It suggested that exposure to an estrogen, like DES, at a critical time in development, could also play a role in the later development of both vaginal cancer in the female and prostate disease, including cancer, in the male. They hypothesized that estrogen exposure during development initiated cellular changes in the prostate, creating "estrogen-sensitive tissue." In effect, they thought, some of the female tissue, from the Müllerian duct, had been retained, and this created permanently estrogen-sensitive tissue which could predispose to prostate disease later in life. The altered ratio of sex hormones during development, they believed, had led to the retention of the female tissue in the males.[4] It is at least interesting that, while McLachlan and his team were investigating events in the womb that might predispose to prostatic and vaginal cancer later in life, in Europe Skakkebaek was piecing together evidence that showed testicular cancer could well have its origins in fetal life. Different diseases, different experiments on different continents, but both teams had arrived at the same worrying conclusions.

In humans, diseases of the prostate are very common. According to Professor Santti, by the age of eighty nearly all men will experience the symptoms of either prostatic cancer or prostatic hyperplasia. Symptoms usually first appear between the age of forty and fifty, although most people do not seek medical help until symptoms are more advanced at around sixty. Prostatic hyperplasia, where the gland swells and presses on the urethra, can be treated with drugs or surgery. However, surgery to remove the prostate cancer is not usually undertaken unless strictly necessary because the nerves which are important for an erection pass by the gland so closely that even very skilled surgeons run a risk of damaging them and causing impotency. Prostatic cancer is slow growing and gives relatively few symptoms which makes it hard to detect. It has become the most common cancer in men. The number of cases reported has doubled in the last ten years in the West. Rates of the disease are estimated to be much higher in the West than in Asia or South America. "Part of this increase is because people are living longer and have more time to develop the disease," Santti points out. "However, according to figures from the Finnish cancer reg-

istry, even if the data is age-adjusted, there is still an increase, although this is disputed."

"I think, based on our research, it's entirely plausible that early
estrogen exposure can play an important role in prostate cancer,
and I think this really cries out to be investigated," argues Mc-
Lachlan. But at the time their study attracted little attention.

All these bizarre changes, and the worrying link to prostate cancer following synthetic estrogen exposure, raised for McLachlan one
intriguing puzzle. What was the mechanism which enabled the synthetic estrogen DES to mimic estrogen and have such a powerful
long-term effect? "That, I believe, is the way you understand science best, by understanding the mechanism," he says. But as they
delved deeper and deeper into these issues, they stumbled across
something else remarkable. "There was an even more fundamental
change in the male mice. The male mice were actually being feminized at the *molecular level*," he found. "The ones who had been
exposed prenatally to DES would actually express female proteins
in their reproductive systems later in life and this is something that
never happens normally in a male mouse."

Lactoferrin is the major secretory protein that is normally made
by genes in the female uterus when exposed to estrogens. Otherwise known as "uterine milk," it was thought to nourish the embryo, or to encourage growth. But when they examined the gel
showing the proteins which were being secreted, they could see
the telltale trace of lactoferrin in the males!

"It was amazing to me, because males never produce female
proteins in their seminal vesicles, which produce the semen. They
still looked like normal seminal vesicles, they still produced semen,
they were still the right shape. But when we stained for lactoferrin,
there it was! The males were producing female proteins. The cells
were living in the male, but behaving as though they were in the
female uterus." It was a form of what McLachlan termed "molecular pseudohermaphroditism." Exposure to the synthetic estrogen
had somehow "imprinted" female genes in the male reproductive
tract in such a way that they could become active, or switched on,
later in life.

"We were very impressed that this gave us new information

about how the reproductive system might develop at the molecular level," recalls McLachlan. "This gave us enormous concern that some estrogens may actually affect the reproductive system in a very profound way, at a very deep level, because the male mouse is expressing female genes at the molecular level." The long-term consequences of this still need to be determined, he warned in his article on these findings in 1989.[5]

The team had now shown that exposure to a potent synthetic estrogen during development could cause three major changes. Firstly, there were the structural changes obvious to the naked eye at birth: the reproductive abnormalities such as male and female reproductive systems developed side by side. Secondly, there were the delayed effects, visible under the microscope: changes in the cells that were consistent with vaginal or prostate cancers. Thirdly, they had shown that the estrogens, at an even more fundamental level, had altered the way genetic material in the cells expressed itself. Messages were now scrambled, so that males could express female proteins.

Not satisfied with this level of analysis, John McLachlan, who was by now beginning to enjoy a considerable reputation, wanted to know *why*. How had the synthetic estrogen caused such fundamental changes? And if DES could do all this, could other molecules mimic estrogen and have the same effect? This puzzle, of how estrogens exerted their effects, was a challenge originally posed by the English scientist Charles Dodds back in 1938, when he made the first synthetic estrogen, DES. Dodds knew DES was a potent estrogen, but had no idea how it exerted its estrogen-like effects.

For years it had been thought that all hormones, including estrogen, exert their effects in the cell by means of a "lock and key" mechanism. The idea is that the hormone molecule fits into a larger molecule called a receptor, rather like a ship docking at a harbor. Once fitted into place, the hormone molecule can trigger the further reactions in the cell. "It's that fit that then turns on the biological activity associated with this hormone," McLachlan thought. "Now the other molecules such as DES that work like estrogens probably work the same way. They would fit similarly in this kind of receptor, fit this same space, and turn on the effect. . . ."

If this were true, it would suggest that the key to the puzzle lies in the molecule's shape. Somehow DES was able to fit into the same receptor and turn on the effect of an estrogen. "In the early seventies we thought we understood it. There did seem to be certain defining characteristics in shape. All estrogens, we thought, must have a benzene ring and an OH group and a certain length of molecule. This would then enable them to fit into the estrogen receptor."

As McLachlan studied this process, and tried to work out why it occurred, to his dismay he came across evidence which added newfound urgency to these inquiries. He realized that another synthetic chemical to which we are all exposed, which can be measured in our blood and our fat, was also weakly estrogenic: the notorious pesticide DDT. Even more perplexing, this molecule had a different shape.

The Legacy of DDT

The story of DDT, which in its day has been both one of the most celebrated and vilified chemicals of the twentieth century, reveals our ignorance and our innocence when exploiting the dazzling new wonders of science. In the mad scramble to benefit—and, of course, to profit—from this extraordinary molecule, we just couldn't anticipate what lay in store. We had no name for it, no means of recognizing it: the problems we were to face were quite simply outside of our experience. Even when warning signs were there, we didn't see them. The fact that we are still finding out new things about its properties, fifty years after contaminating the globe with it, highlights the difficulties.

From the perspective of science in the 1930s and 1940s, it's easy to see with hindsight just how this state of affairs arose. In the early part of the century, modern science, like a conjurer, had produced magic out of a hat. Ordinary lives were transformed more dramatically than perhaps in any other era. Electricity brought light at the press of a button. Penicillin, available in the doctor's surgery, snatched infants and children back from what would have been certain death a generation earlier. Science could do no wrong and scientists themselves were heroes.

In the 1930s there were a number of developments in a branch of chemistry known as organic chemistry. This is the chemistry of the carbon molecule, which is the basic building block of life, hence the term "organic." It is so versatile that it can be put together in an infinite number of ways, depending on how the carbon atoms are joined together, in chains or rings, to make the molecule. The number of configurations that could be created seemed almost limitless. It could be argued this was the heyday of organic chemistry, so many totally new compounds were synthesized and developed by industry to create amazing products. This was an era of optimism and confidence in science, when people would look forward to whatever might emerge next from the laboratory. Science appeared to be a great provider; whatever emerged it was bound to be good.[6] New substances created each week were to have a dramatic effect on our lives and their manufacture was to give rise to today's chemical giants.

DDT itself was first synthesized in 1873 by a German scientist, who described the colorless, crystalline solid, but apparently failed to realize the chemical's potential. It was rediscovered in the 1930s by Paul Hermann Müller, a Swiss chemist, who was trying to create the ideal insecticide. He believed that it would be possible to design a safe insecticide, since, he reasoned, the absorption of toxic substances by insects was entirely different physiologically from their absorption by mammals. In his laboratories at the dye factory of J. R. Geigy in Basel, Switzerland, he tested many substances. In September 1939, just as Hitler was marching across Europe, Müller stumbled on the perfect solution. He found that dichlorodiphenyltrichloroethane, DDT, killed flies fast, destroying the insects nervous system in just a few moments. It was soon clear it had rapid toxic action with all insects and was heralded as a breakthrough.

During the war it was very much in demand. Applied as a dust or by spraying the aqueous suspension, it was effective against lice, fleas and mosquitoes, the carriers of killer diseases, typhus, plague, malaria and yellow fever. When Naples was liberated by the Allies, the entire population was dusted with DDT to wipe out an epidemic of typhus. Soldiers being recruited were treated with it. Shirts were impregnated with it. Blankets dusted in it. In the tropics it was used all the time. It was celebrated as the new war weapon of the Allies because of its ability to destroy killer pests.[7,8]

At first DDT seemed like yet another miracle of modern science. Agriculture was plagued with insects. Whole crops used to be wiped out and insects were perceived as a threat, not just to us, but to the "very health of the land."[9] DDT promised victory in the war against insects. It was even used liberally in the home, in sprays or aerosols called "insect bombs." Advertisements of the fifties reminded people of a past with insects, of picnics invaded by swarms of ants and wasps, of entire crops withered and shrunken by pests. It was thought to be so successful and so safe that there was even a drink, a Micky Slim, gin with a spot of DDT, which was supposed to give a little kick.[10] It was the most widely used insecticide for more than twenty years and a major factor in increasing world food production and the suppression of insect-borne diseases. Some passionate advocates credited DDT and related insecticides with saving more than 100 million lives throughout the world.[11] As if in answer to a prayer, this extraordinary chemical held out promise of obliterating some horrific diseases in the Third World and ensuring a much needed increase in food supply. In 1948, Paul Müller was awarded a Nobel prize for physiology or medicine for discovering the potent toxic effects on insects of DDT.

As the cameras flashed and the applause swept the floor in Stockholm as Müller collected his prize, quietly in New York two rather less well known scientists were making a most strange observation about DDT. Frank Lindeman and Howard Burlington at Syracuse University were trying to understand if DDT had effects in animals other than insects. Could mammals and birds be at risk? To find out, cockerels were given a small dose of DDT for sixty days. To their surprise, they found marked adverse effects. Male animals exposed to DDT in their laboratory did not develop secondary sex characteristics as they should. Combs and wattles, skin folds around their necks, for instance, were less than half the size of those belonging to creatures in the control group. The testes were severely underdeveloped, on average five times smaller. From this they deduced: "It seems therefore that the possibility of an estrogenic action of DDT is at least worthy of consideration. . . . These findings suggest DDT may exert an *estrogen-like action* in white Leghorn cockerels." They even speculated about the degree of similarity in the molecular shape between DDT and synthetic estrogens like DES. This was the first warning of delayed and more subtle effects.[12]

But these odd findings were published in 1950 and soon forgotten, buried by years of scientific papers before their significance would be understood, even though in 1957 these results were reinforced by the work of Richard Welch and Ronald Kuntzman at Burroughs Wellcome Research Laboratories. They reported that DDT could alter the formation of certain enzymes in the liver, and this in turn altered the formation and regulation of the sex hormones estrogen, progesterone and testosterone. They even speculated that this effect on sex hormone metabolism could help to explain the decrease in population of certain species of birds.[13] Later studies also confirmed that DDT can alter hormone metabolism, and thereby affect reproduction.[14]

Meanwhile, the 1950s were to see record sales for DDT. To meet the demand, there was an almost fivefold increase during the decade in synthetic pesticides production, with a value of over a quarter of a billion dollars.[15] Archive footage of the time shows how casually it was used. Children were sprayed with it. Suburban lawns were doused in it. The white spray from helicopters would be seen in all agricultural districts. It was believed to be so safe and effective that one entomologist even claimed you could eat it and, to prove the point, would swallow a pinch or two.[16] After all, it was very clear that millions of troops had survived regular spraying. If this was a death substance, death was a long while coming.

But in 1962, the honeymoon came to an end. Rachel Carson, a biologist from the U.S. Fish and Wildlife Service, published *Silent Spring*. Her book was an immediate sensation and is largely credited with starting the environmental movement that became so popular in the seventies and which prompted the demise of DDT in the

West. She described changes in wildlife that, she argued, were plainly visible to see and hear at the time, even in a suburban backyard: birds trembling violently and then dying; the absence of the sounds of spring, such as droning bees and singing birds; browned and withered vegetation; animals without young; even, she claimed, children suddenly dying without explanation. "What had silenced the voices of spring in countless towns across America?" she demanded. "No witchcraft, no enemy action had silenced the rebirth of new life in this stricken world. The people had done it to themselves." They had done it, she argued eloquently, with the use of synthetic pesticides, "the elixirs of death," and foremost among them was DDT.[17]

At the time she was told she was mistaken and her approach was unscientific. The industry accepted that DDT was harmful to insects, but believed it could not harm human beings. But the public thought otherwise. Rachel Carson was feted. She found herself invited to the White House, where President Kennedy himself had been quizzed about her book; she was celebrated on TV and serialized in the *New Yorker*. Somehow she managed to keep up the battle with her industrial critics, although she was herself engaged in her own private battle, with breast cancer, which was to take her own life the very next year.

Her work triggered a wave of new studies. Yet the spraying continued. When the inventor of the chemical, Paul Müller, died on 12 October 1965, the case against his discovery was mounting. At the height of the spraying there were extensive die-offs of various species of wildlife, including top predators, such as falcons, eagles, pelicans, salmon and bass, many of which were shown to have high levels of DDT in their body fat.[18] Reports discussed cancer, tumors and strange reproductive effects.

The final challenge came quietly in the midwestern state of Wisconsin in 1968. Two organizations, the Citizens' Natural Resources Association and the Izaak Walton League, requested the state to determine whether or not DDT was polluting the state's waters. A little-known law in this state permitted anyone a legal hearing if they thought they could prove water pollution in the state. This was in effect a direct challenge to the state's right to continue to use the pesticide at risk of contaminating the waters. It was to become "the trial of DDT."[19]

It was soon clear if industry lost in the state of Wisconsin, others could follow suit. Big guns lined up on either side. The protesters were supported by the Environmental Defense Fund and many scientists worldwide. Industry set up their own special panel of experts, at least one of whom was prepared to claim that DDT was entirely harmless, in spite of mounting evidence to the contrary.[20]

What clinched the court case was new Swedish evidence from the chemist Goering Lefors showing that DDT had been identified in human breast milk! The Environmental Defense Fund lawyers called him and urged him to give evidence at the trial. At the same time they alerted all the newspapers, from the *New York Times* to the *Chicago Tribune*. Lefors explained in great detail how DDT accumulates up the food chain, and can be found in milk, fat and even brain tissue. His testimony stole headlines across the country and won the day for the environmentalists. DDT was banned.[21]

With the same enthusiasm that reporters had seized on the successes of science in the thirties and forties, in the seventies the tide turned. Now every last detail of injury or deformity that may have been caused by chemicals became the subject of detailed media scrutiny. It was soon realized, as Rachel Carson had argued, that the very properties of the molecule which make it such a successful insecticide could also make it harmful to other species.

Firstly, its *persistence*: DDT is not readily destroyed. The molecule has a half-life of about 100 years, which enables it to remain in the environment long after spraying.[22] It also persists in the bodies of animals which have consumed it, where it may be converted into the related compounds DDD and DDE, the main "metabolites" or breakdown products of DDT in the body, which can be retained in tissues for years. Mostly DDT and its metabolites are stored in our fat supplies, because, like so many synthetic estrogens, they are fat-loving or "lipophilic." In plants and animals it will be readily assimilated in their lipid or fat stores.

Its persistence enables it also to be concentrated as it goes up the food chain, a process known as *bioaccumulation*, or biomagnification. For instance, tiny organisms in the sea, called plankton, will absorb a small amount from sediments and water. This could be in trace amounts of 1 part per million (ppm). Fish that live off the plankton will, over time, build up higher concentrations in their tissues, to perhaps 10 ppm. This process continues up the food chain, with predators always accumulating more than the species they consume. As man feeds off the predators, he can accumulate much higher levels. In some populations it has been found at levels of over 30 ppm.

Not only is it persistent and bioaccumulative, but it can also be transferred directly *across generations,* with one generation passing on their chemical load to the next. In humans this direct transfer can happen across the placenta and through breast-feeding. On average, DDT is now found in breast milk at levels of 30 parts per billion (ppb) of whole milk. In milk fat, levels are higher, around 1 ppm. Typically in developing countries where DDT is now more extensively used, levels can be 100 times higher than this. One of the highest levels ever reported, of more than 100 ppm DDT in milk fat, was from human milk in Guatemala in the 1970s due to indoor spraying with DDT to try to reduce malaria. When breast-feeding, a mother runs down her fat reserve. In doing so, she may literally pass into her baby a significant part of the toxic load accumulated in her fatty tissue over her lifetime.[23] For a fuller discussion on chemicals in human milk, see Chapter 8.

Apart from being persistent, bioaccumulative, and transferred across generations, it was soon found to be incredibly widespread. By 1970, telltale traces of DDT and DDE had been found in the remotest regions of the Arctic, in the fat of Arctic wildlife, seals and penguins, in the Antarctic snow, and even high up in the air, as high as 21,000 feet over India.[24] In a study of seals and porpoises off the coast of Scotland, levels of DDT were found to be as high as 73 ppm. The authors Holden and Marsden warned in 1967: "The degree of accumulation of pesticide residues in an environment not deliberately contaminated, and in a species far distant from the target organisms of persistent pesticides, underlines the impossibility of confining such chemicals to the areas of application."[25] It was becoming clear that DDT is spread around the world by spraying, which contaminates air and water; by accumulation up the food chain in the sea and on land; by the export of crops and animals; and also by direct transfer across human generations.

Once the tide had turned, evidence of the harm it could do emerged fast. Several laboratory studies had shown since 1947 that DDT could act as a carcinogen, producing tumors in laboratory animals. There was even such a study from the American Food and Drug Administration, but it was ignored.[26] By 1968, evidence emerged that DDT may be carcinogenic to humans. Radomski and his team at the University of Miami compared levels of DDT in people who died of cancer compared with people who died in accidents. They found that those who died of cancer of the lungs, stomach, rectum, pancreas, prostate and bladder had "remarkably high" levels of all pesticides tested, "averaging two to three times the normal concentration and twice the levels of DDT." Accident victims averaged just below 10 ppm of DDD, DDE and DDT, whereas those with cancers averaged between 20 and 25 ppm.[27] As the evidence of a link between DDT and cancer emerged, the American Environmental Protection Agency finally restricted the use of DDT.

DDT has now been banned or restricted in the West for twenty years or more in most countries and the levels in our bodies have fallen. However, because of its widespread use in developing countries, especially in malarial control, it is estimated that the worldwide use of DDT almost certainly exceeds levels that were used historically in the West.[28] In 1991, the United States still exported nearly 100 tons of DDT.[29] According to World Health Organization figures, Mexico and Brazil each used nearly 1,000 tons of DDT in 1992.[30] Besides affecting the health of the local populations, such use will continue to add to the burden of contamination around the world.

"It was really surprising to us initially when the first reports of DDT being an estrogen came out," recalls McLachlan. "To find out that a chemical that is so widespread in the environment can do this was a great concern." What was even more surprising when he checked the literature was to find that DDT had been reported to be estrogenic in the 1950s and yet we did not recognize the significance of this for some time.

Soon after this, he came across reports indicating that other pesticides, too, might act like weak estrogens.[31] Even more worrying, there was evidence that some of these might affect reproduction in humans. In 1975, at the now closed plant of Life Science Products Company in Virginia, workers were exposed to the pesticide kepone and were found to have high levels of the substance in their blood and tissues. The case caused a considerable stir, not least because a team of researchers from the Medical College of Virginia, headed by Dr. Guzelian, found that fourteen of the male workers were now probably sterile. Semen analysis showed markedly abnormal sperm development and reduced sperm motility. The team also showed other severe problems, including tremors, headaches and memory disturbance.[32] The estrogenicity of kepone was speculated upon.

"Kepone challenged our whole thinking about how hormones work," recalls McLachlan, "because it was such an unusual chemical. It is a totally different shape to the other estrogens. It is a cubic molecule with chlorines sticking out. It was just intriguing.

"What is more, kepone brought home to us that these chemicals could affect human health. Up until then we had thought this was a wildlife issue. But at this point we really started to question which chemicals could act like estrogens and what effects they might have." Literature searches suggested the list of possible estrogenic pesticides was growing. They included DDE, the breakdown product of DDT, kelthane, heptachlor, kepone and methoxychlor. It's impossible to predict from their shape which molecules act like estrogens. Many different molecular shapes have turned out to be estrogenic.

Puzzled at the diversity of shapes, he contacted a leading X-ray crystallographer, William Duax, who was working on an atlas of hormones at the Medical Foundation in Buffalo. McLachlan invited him to figure out what it is that makes all these different molecular shapes act like estrogens.

"Sure, I'll look into it," he replied, confident that this puzzle could be solved quickly. Samples of all chemicals that might act like weak estrogens were duly sent to the Medical Foundation. McLachlan, in anticipation of having an answer to the mystery, organized a conference in 1975 and invited William Duax to give the first talk.

"We were all thrilled. Bill was going to give a talk, solving the structure of estrogens. He'd been working on it for weeks and he was the best in the world." But when Duax began his talk, he was forced to admit that the task was complicated by uncertainties and that they had not been able to crack the puzzle.[33] Perhaps if they had another meeting in a few years' time, they might have a solution.

"But to this day," reports McLachlan, "we still don't know what makes an estrogen, or an anti-estrogen. It's a puzzle we have yet to solve, the structural basis of estrogenicity, and it is a very important problem."

However, it was becoming clear that these chemicals may be disrupting the action of natural estrogen in different ways. "What these synthetic estrogens can do," he explains, "is firstly fit the receptor and *switch on* the hormonal effect just as the natural hormone does. Or they may *block* the receptor by fitting in the wrong way and keeping the normal estrogen from going in, making a hormone block. Thirdly, they can *turn on more receptors* and that can make the estrogen action more potent. Finally, they can even *change the metabolism* of the natural estrogen in such a way that it ends up in a different form in your body and this can have an influence on the action of the estrogen in an indirect way."

In other words, these molecules have the extraordinary power to undermine the fundamentals of sex hormone chemistry. So why, he wondered, are the receptors—the lock into which our natural estrogen fits like a key—not more specific about which keys they accept? Why can any number of molecules fit? The answer, he felt, stems from our evolutionary history.

It has taken millions of years for our internal environment to evolve. All the extraordinarily intricate details of our cell physiology and biochemistry developed over millennia. The estrogen receptor system is similar in all mammals and is not very specific simply because until recently it didn't need to be.

It was only with the astonishing developments in organic chemistry this century that suddenly hundreds of new organic chemicals, based on carbon, the building blocks of life, were created and released into the environment. Many of these substances our bodies simply hadn't seen before. We had no mechanism within us for recognizing them or dealing with them. How strange it seemed that somehow we had created substances, which we had sprayed worldwide, that had the potential to mimic our own sex hormones and interfere with our sexuality! No one even dreaming up the plot of a science fiction novel could envisage such a bizarre twist to our fate.

"These molecules could be having subtle effects in many different species which now share this common synthetic environment, and with many different diseases that are hormonally triggered," he thought. "We need to understand the increasingly complex interaction between our synthetic environment and our internal environment."

He became so concerned that, during the 1980s, he continued checking out other different families of chemicals to see if they were also hormone mimics, and if any of these would give insights into the mechanism. Every year more and more chemicals were identified as estrogenic. To his dismay and fascination he found that yet another ubiquitous group of organochlorines, known as polychlorinated biphenyls, or PCBs, which are now widely dispersed in the environment, also appeared to be having reproductive effects.[34] Some of the PCBs appeared to have a structure very similar to that of DES. Could these also act like estrogens, he wondered.

The Legacy of PCBs

PCBs were first discovered in the late nineteenth century, but, like so many of the other organochlorines, were not developed commercially until the 1930s. In 1929, in the labs of the Swan Chemical Corporation in America, it was confirmed that by mixing biphenyl with chlorine it was possible to synthesize new compounds, which became known as polychlorinated biphenyls, or PCBs. Depending on the proportion of chlorine to biphenyl, different types of molecules can be made. In fact 209 different types of PCB have been formulated.[35]

This colorless, viscous fluid looked very ordinary, but was soon found to have some extraordinarily bankable properties and they were used extensively during the Second World War. In 1935, Swan was taken over by Monsanto Chemical Company, and they continued to develop commercial uses for PCBs in America. In Europe and Japan they were made by other manufacturers.[36] Unusually high chemical stability, considerable heat resistance, low flammability and high electrical resistance all led rapidly to a range of industrial applications worldwide in the fast-growing electrical industry. With the data available in the 1930s it would have been difficult to envisage the final outcome of the story of this molecule. It was used in transformers, electromagnets, switches, circuit breakers and voltage regulators, and to improve electrical insulation. By the 1980s over 40 percent of all electrical equipment in the U.S. contained PCBs. PCBs were also used in heat transfer systems, as additives in lubricating and cutting oils, as fire retardants, plasticizers, and additives to cement, plasters, printing inks and hydraulic fluids, even brake fluid in cars. They were the molecule of choice for regulators and manufacturers alike because they seemed so safe. They were so much in demand that they were available in containers ranging from 50-pound to 600-pound steel drums and by the tank and carload.[37]

The enthusiasm for this family of molecules changed dramatically as a result of an accidental discovery. In the 1960s, scientists investigating DDT levels in the wake of Rachel Carson's polemic began to find unexpected peaks in the tracings of their instruments, which couldn't be explained by the usual pesticides. The mystery substance continued to be elusive, and thought to be of little consequence, until 1966, when a Swedish analytical chemist from the University of Stockholm, Søren Jensen, identified PCBs in fish. He analyzed over two hundred pike from different parts of Sweden. Every one of them produced the telltale peaks of PCBs.[38]

Worried by these results, he continued his research and found PCB in virtually every living thing he examined. It was in fish spawn. It was in an eagle they retrieved from the Stockholm ar-

chipelago. It was even in eagle feathers they obtained from the
Swedish National Museum of Natural History, or at least those
collected since 1944. It was in conifer needles. Then he found it
in his own hair! It was in his wife's hair, and even in hair from his
five-month-old baby daughter! He presumed she received her dose
through the mother's milk. In 1966, in the *New Scientist,* these
findings were outlined in "A report of a new chemical hazard."
He described how the substance had also been traced in the air
over London and Hamburg and even in seals found off the coast
of Scotland. "It can therefore be presumed to be widespread
throughout the world," he concluded, and "may enter the body
directly through the skin, by breathing, or by way of food, especially fish."[39]

It was not long before other scientists confirmed his findings. In
1967, Monte Kirven at the San Diego Natural History Museum
was studying the reasons for the disappearance of the peregrine
falcon. In one abandoned egg he had found high levels of DDE
and also some other unidentified compounds. When Jensen's work
was published, Kirven went back and retested the mysterious compounds. He found they were indeed polychlorinated biphenyls.[40]
Soon it was realized that PCBs, just like DDT, had become widespread contaminants, not just in industrialized countries, but reaching remote parts of the Arctic and the Antarctic. Sometimes the
amounts found were high, as large as 1,980 ppm in the fat of peregrine falcons in North America, and even a staggering 17,000 ppm
in the fat of the white-tailed eagle in Sweden.[41]

But how had the PCBs become so widely distributed in the
environment when, unlike DDT, they were not sprayed intentionally on the land? PCBs would leave the factories in sealed drums
and were largely used in closed systems. It began to emerge that
gradual wear of PCB-containing products could result in their slow
release into the atmosphere as vapor. Many products containing
PCBs end up in city dumps where they may be burned, a process
which may also release PCBs into the air. In addition it can be
identified in industrial wastes which may be discharged into the
environment. Amazingly, without intentionally spraying the product, we had somehow scattered it across the globe.[42]

Just as with DDT, once identified, the dangers of PCBs began to emerge rapidly. In one study, Dr. David Peakall, at Cornell University, exposed birds to either DDT or certain PCBs. He found that PCBs, like DDT, could increase enzyme activity in the liver, which in turn affected the metabolism of the sex hormones, increasing the breakdown of estrogen. In his study, PCB was even more effective than DDT in increasing the breakdown of estradiol, which he believed would ultimately make the animals sterile, by altering the ratios of sex hormones.[43] Could the presence of PCBs in conjunction with pesticides account for the rapid decrease in some birds of prey, he speculated.

Despite such reports, over a decade was to elapse after Jensen's warning before production was stopped in most countries. Some estimates suggest the over 1.2 million tons have been manufactured worldwide.[44] Since PCBs are very persistent and do not degrade easily, it is virtually impossible to remove them from the environment. Indeed, in a recent study, a Canadian scientific team, sampling cores from the Agassiz Ice Cap in the High Arctic, found that levels of PCBs deposited in the snow and ice had changed little in twenty-five years. Working on the frozen snowy wastes of Ellesmere Island, they moved 20 tons of snow, making a pit over 8 meters deep. Ice cores taken at different levels can provide a record of chemical deposition over the years. They found 848 nanograms per square meter of PCBs were deposited in 1989–90, which was not far removed from the levels deposited nearly twenty-five years ago, when PCB production was still ongoing: 930 nanograms per square meter in 1967–68.[45]

Because PCBs are very persistent, they are present in trace amounts, parts per billion, in all surface soils, the air, vegetation and water. Like DDT they accumulate up the food chain, and are found in higher concentrations in man and marine mammals. They can be transported through the air as vapor and can enter the human body through the lungs and the skin and, perhaps most importantly, through our food.[46] They are typically found in the fatty, oily products, such as fish, dairy and some meat products. Ironically there is evidence that some of the more toxic PCBs are the more readily accumulated.

The human fetus is exposed in the womb, where PCBs can be transferred across the placenta. This exposure continues after birth, especially during breast-feeding. Allan Jensen, in one review of studies, identified high levels of PCBs in human breast milk: in Germany, levels were over 100 ppb; in Denmark, they reached 140 ppb; and in Spain, they went up to 250 ppb. "The relatively high levels in breast milk result in very high daily intakes of such chemicals by breast-fed infants," he warned, "and are a potential hazard to this risk group."[47] Like DDT, most PCBs are not readily soluble in water, but they are soluble, and tend to accumulate, in fat, and can be found in the fat and milk samples of practically all people from industrialized countries.[48] PCBs from the mother's fat store are effectively mobilized and transferred into the baby directly during breast-feeding. Estimates suggest that the World Health Organization recommended limits may often be exceeded.[49] Jensen is not the only scientist concerned at the levels of PCBs in breast milk. A team from Reading University, led by Professor Raymond Dils, recently found that breast milk from mothers in the Reading area had a concentration of PCBs which was at least twenty times higher than levels shown to cause adverse effects in humans. Sixty different types of PCBs were found in breast milk, and some of these are thought to promote carcinogenesis in laboratory animals.[50]

Concerned at data accumulated so far, in the mid-1980s John McLachlan's team began studying PCBs. They were interested to find out if some PCBs had weak estrogenic activity. Intriguing strands of evidence hinted that this family of synthetic chemicals could have reproductive effects in fish and affect fertility in other wildlife. Early laboratory work had also suggested estrogen action, in that certain PCBs could make uteri grow, a typical estrogenic effect.

McLachlan's team obtained a number of different PCBs from commercial sources. These were tested for their ability to bind with estrogen receptors. In addition they were tested for their ability to increase uterine weight. They found that some PCBs were indeed weakly estrogenic. "We actually were surprised at how estrogenic some of them were," says McLachlan, "one-fortieth the potency of estradiol, so it was quite a marked effect that put them in the same range as the other weak estrogens." In their paper on these findings they warned that their results may "reasonably account for . . . unexplained episodes of estrogenic activity in human populations exposed to complex mixtures of PCB residues in the environment."[51]

By chance, soon after this, David Crews, a biologist from the University of Texas, came to visit McLachlan's team in North Carolina. He was invited to give a seminar on a unique study that he had just undertaken with turtle eggs. Sex development in turtles is temperature-dependent. If eggs are incubated at 26°C, they will all be males. But Crews reported that if he painted the eggs with 10 micrograms of the female hormone estradiol and incubated them at the usual temperature for males, they would all be female. Exposure to the female hormone had reversed the sex. McLachlan approached him at the end of the meeting and explained their findings on the estrogenicity of PCBs. He asked if Crews was interested in testing out PCBs to see if they were sufficiently estrogenic to reverse the sex development in turtles. "That would be great. That would be a real interesting study. Let's do it!" was the enthusiastic response.

Thirteen coded solutions of PCBs were duly sent down to the laboratory at the University of Texas. Crews soon called with remarkable news. Two of the PCB mixtures were particularly potent. They had reversed the sex development of many males at 26°C. When McLachlan checked the codes, these were the two most estrogenic PCBs: chlorinated 3- and 4-hydroxybiphenols. He suggested they tried painting them on the eggs together. This time the result was striking. Acting together, they were twenty times as potent as acting singly.[52] "This was a strange result," recalls McLachlan. "It was difficult to explain. But we noted it. It was there to see. People commented on it. We thought it would be important."

By now, in 1993, McLachlan was the director of the laboratories he had entered as a young man, the National Institute for Environmental Health Studies. He had lost none of his enthusiasm for science, nor his fascination with the details of development that had brought him here over twenty years previously. In this time he had built up a considerable reputation within his speciality. From his large office on the top floor he had a view beyond the lake to the forests and the far horizon. The lake was still there. The flag was still flying. But the world had changed, because now we took nothing for granted. There was a much greater awareness that the womb, far from being inviolate, was a fragile thing, that development of a human being is a process that is exquisitely sensitive even to subtle changes in hormonal milieu. McLachlan's intuition that had brought him here to study DES in the late sixties had turned out to be frighteningly correct.

But his warnings about possible harmful effects of estrogens in the environment had not been heeded. As early as 1979, he organized a symposium called "Estrogens in the Environment," where the extent to which we had accidentally distributed chemicals in the environment that could act like estrogens was discussed, including the widespread use of DES in the cattle industry and DDT in agriculture. "Thousands of chemicals are introduced into our environment with little knowledge of their effects on two physiological processes which are central to our survival as a species: reproduction and development . . ." he had warned in 1979.[53] But at the time there were few, if any, documented human health effects from environmental estrogens, and beyond a small circle of scientists no one paid any attention.

McLachlan himself felt it was significant, and in 1985 he organized a second conference on estrogens in the environment at NIEHS, this time including data on precocious puberty in young children. Over the years he had written several books and had presented his ideas at international conferences. Although his work was appreciated within narrow scientific circles, where he was admired as a pioneer of a new approach to environmental hazard and as the "father" of the field of environmental estrogens, the work did not hit the headlines. He published his ideas in the leading scientific journals and, like many scientists, never courted the popular press. The press, in turn, appeared oblivious of the wider significance of these studies.

But McLachlan himself was quite certain, as he had been for fifteen years, that this would be important. "There are so many complex processes in our biology and physiology that to imagine a sea of estrogens which can affect our development, reproduction and pathology in a way that is still unknown, to me has profound implications for the future—not only for a variety of different species, but also for humans."

It took the announcement of a fall in human sperm counts for his ideas to be catapulted into the spotlight.

In Finland, Professor Risto Santti, who had worked with McLachlan on DES and prostate cancer, also had a very successful career. He had served as chairman of the Finnish Medical Research Council and as vice-dean of the medical faculty at the University of Turku. His consuming interest throughout his career was understanding the effects of the synthetic estrogen DES on the male reproductive tract and especially on the prostate. He had published many papers, given talks at international conferences and pursued his subject with total dedication.

Shortly after his father's death in 1993 he received a call from his mother with some unexpected information. She had been going through the letters her husband had written to her during the Second World War, when he was serving in the Finnish army. In among them were prescriptions written by a leading Helsinki gynecologist, which she had taken when pregnant in 1942.

"Are you interested in the drugs I used when I was pregnant with you?" she asked.

"Most definitely," was the reply. He imagined it would be some innocuous painkiller. She began to read out the prescriptions. But she couldn't decipher the scientific terminology. So she spelled out the words instead. It is impossible to describe his feelings of cold, hard shock and surprise. She could not have spelled out more clearly the kind of future that he believed lay ahead. The drug she had taken while carrying him was the very same drug he had been studying all his life: the infamous DES.

Even for a scientist with no belief in fate or any other irrational force, this was too much of a coincidence to be true. At first, he didn't believe it. His mother had no idea of the impact of her call. She did not follow all the details of his scientific research and he did not enlighten her. As requested, she put the prescriptions in the post to him.

Once the yellowed slips of paper arrived, it was impossible to pretend he had imagined the whole thing. There they were, well-marked with time, the few words in a hurried scrawl that could ruin his peace of mind. What is more, he could see she had been prescribed the drug throughout the critical stage of the pregnancy. His only hope was that for some reason she hadn't taken the drug. Since she was well into her eighties, not wishing to trouble her with his anxiety, he did not ask. He was only too well aware, from years and years of his own research, of the possible link between DES and later prostate cancer, so he went instead for a cancer test. A simple blood test can reveal whether there are antigens present indicating early signs of prostatic cancer.

Although it was too early to detect the tumor, the result was positive. "Do you wish us to do further tests?" asked his doctor. Risto Santti shook his head. He was convinced that what would happen now was preprogrammed by events during the war before his birth. Prostatic cancer typically grew slowly. They could of course treat the symptoms as they emerged. But he knew from years of study that there was nothing he could do to stop the development of the disease. The cancer would take its own course. As a doctor, even with his intimate knowledge of every stage of the disease, he was powerless to help himself.

CHAPTER FOUR

First Suspects

Edinburgh, Scotland, 1992

At the Medical Research Council laboratories in Edinburgh, the coffee room is on the first floor, with large windows overlooking the park. In the style of many truly British scientific institutions, it is not furnished with lavish attention to detail. Clearly, the "money no object" syndrome is not in operation here. On a counter at one end, two huge vats of boiling water allow scientists to help themselves to drinks during the day. The harsh glare of the daylight pours in through uneven blinds. In the distance, the sound of building works and ambulances are an insistent reminder of the hospital next door.

Despite an atmosphere that was hardly conducive to prolonged conversation, by 1992 animated discussion about the estrogens theory could be heard in the coffee room, in the corridors and in the labs. This was not exactly an official project. It was a set of ideas that had come together almost by accident. But for those in the know, it was compelling. The race was on to track down sources of estrogen exposure and to work out how this had changed in the last fifty years. What was it about modern living that meant exposure to estrogens had increased?

Professor Alan McNeilly recalls the concerns: "The implications of this are that there are compounds out there that could affect us. If they can affect fertility, they may well affect other aspects of metabolism. The effects could be enormous because estrogens are involved in lots of different functions in the body. It might affect

my children, my children's children, and so on. That's really serious
if it's a compound that we don't know and haven't detected prop-
erly. We don't want estrogenic compounds out there, uncontrolled
and unidentified. It could cause all sorts of harm, so we have to
get to the bottom of this. . . ."[1]

Dr. Richard Sharpe, aware of the interest, and some skepticism,
raised by the hypothesis, was keen to find the stronger proof they
needed. The ideas were never far from his mind. "I was on my
way to a meeting, catching up with the literature," he recalls,
"when something I was reading jogged my memory about a manuscript which I'd edited sometime ago." He traced the manuscript.
It was Professor John McLachlan and Dr. Risto Santti's work on
prostate cancer and DES exposure. The article described graphically
how exposure to synthetic estrogens could result in changes to the
prostate which might predispose to prostatic cancer in later life.
"This suddenly introduced another new dimension to the whole
story," he thought. "In some respects, the idea that prostate cancer
might be related to estrogen exposure would dwarf everything that
we were interested in because so many people are affected by this
disease."[2] British figures alone suggest the number of cases of prostate cancer reported has almost doubled in the last decade.

But, for Sharpe, there was another uncanny aspect to the prostate
studies.

McLachlan had written in his 1990 article: "For years it has been
assumed that estrogens may be responsible in some way for the
development of prostatic cancer . . . because of the direct effects of
the hormone observable in patients. . . . We suggest an alternative
or complementary hypothesis for a role of estrogens as a predisposing factor for prostatic tumors in which *the critical time for estrogen
action is early in life,* when estrogens are apparently most influential
on the prostate. The critical time for estrogen exposure would pre-
cede, sometimes by many years, clinical appearance of the disease. . . ."[3]

Quite independently they had been thinking along the same
lines, but with different diseases and different sets of data! "The
exciting thing about these findings was how McLachlan's team
were proposing that estrogens brought about these prostatic

changes which might predispose to prostatic cancer," recalls Sharpe. "It was exactly the same mechanism that we were proposing with the failure of testicular descent and testicular cancer. So, again, everything was tied down to a time in life, a point in development when we know estrogens could induce a range of abnormal effects. It also started rumblings in my mind as to whether or not we were looking at something which had more widespread implications than just male reproduction and fertility. Certainly it gives you greater confidence. The more pieces of information that fit the jigsaw, then the more sure you become that you are working on something which could be real and not just a figment of your imagination."

By now articles on sources of estrogens were constantly arriving at his office. "To my, I suppose, great pleasure, one of the first articles I picked up described how humans now live in a veritable 'sea of estrogens,'" says Sharpe, "that exposure is far greater than was the case forty or fifty years ago and that increased exposure comes by a variety of routes." There were articles on dietary changes, synthetic estrogens used to fatten livestock, such as DES in the 1950s, and the pill and pharmaceuticals. Working on the hypothesis that estrogens were affecting Sertoli cells, which in boys can continue to develop until puberty, they had to check out all the routes of estrogen exposure that had changed in the last fifty years, both in the womb and through childhood. By chance, they hit a rather surprising source.

As Sharpe was leaving work one Friday night, he ran into Professor Alan McNeilly, an expert on lactation and fertility. They discussed developments in the estrogens hypothesis as they made their way to the parking lot. "As I was getting into my car, it suddenly just struck me," recalls Professor McNeilly, "of course, another major source of transfer of estrogens into the baby is through milk, the mother's milk, and more importantly bottled milk. The decline in breast-feeding since the war means that most babies in the Western world are taking formula milk at some point, and many even from birth. So I got out the car, and shouted across to Richard, 'Hey Rich, what do you think about transfer in milk, because that would get directly into the baby? I think we ought to look at that, you know.' "

"We both realized immediately what he was saying," says Sharpe. "Most bottle feeds are made up out of cow's milk. Was there something different about estrogens in cow's milk?"

They soon found there *was* something different about cow's milk since the war. The levels of natural estrogens are much higher. Improvements in the efficiency of dairying have enabled milk to be taken from cows that are already in calf. Previously, when cows were bearing a calf, they were much less likely to be producing milk as well. However, by timing insemination correctly, usually three months after the birth of the last calf, it is possible to achieve a high milk yield for most of the year and also obtain offspring. The dairy cow is the only farm animal which is required to be lactating and pregnant simultaneously. This practice has been criticized by some experts in animal welfare, such as Professor John Webster from Bristol University, who describes the cow "as the hardest worked of all our farm animals," and has linked several diseases, including lameness and mastitis, to the high yield of the cow.[4]

However, during pregnancy, cows, like all mammals, produce much *higher* concentrations of their own natural estrogens. Since cow's milk is used to make most formula feeds for babies, it seemed entirely possible that babies were being inadvertently dosed with extra estrogens in this way. "That happened on a Friday and I spent all weekend worrying about it," Sharpe remembers, "because I thought if it were true then this would create the scare of a lifetime and would have tremendous implications."

The obvious test was to take formula milk and measure the levels of estrogens. In discussions over the weekend, they realized that there were indeed scientists, such as Professor Brian Heap in Cambridge, who were studying estrogens in cow's milk. As soon as he arrived at the office on Monday morning, Richard Sharpe telephoned the Cambridge laboratories. "Brian Heap told me, yes, for certain there were estrogens in cow's milk, but—and thankfully, I think—that when they measured estrogen levels in formulated baby milk, these estrogens had disappeared. So it looked as though it wasn't the switch to bottled milk that was responsible in this case. I breathed a huge sigh of relief. It would have been nice to have a neat explanation, but at the same time, it would have been a very worrying explanation to have."

"But in many ways this was worse," says McNeilly, "because if the changes weren't due to something specific like changes in milk production, then what were they due to? It seemed to have to be a different sort of compound that we hadn't thought about, or didn't know about. Maybe we couldn't even measure it!"

However, the case on milk was not quite closed. Richard Sharpe is still concerned about what does happen to these estrogens in milk. "We don't know if it does get into our bodies, if we absorb it, whether it might be influenced by other factors. Nor do we know why the estrogen disappears when you make formulated milk and where that estrogen goes, whether it might end up somewhere else. At the moment we just don't know. . . ."

But the team were soon alerted to another important source of estrogens. Every year in Edinburgh there is a conference, known as the Simpson Symposium, held in honor of Sir James Young Simpson, who 150 years previously had made major advances in the field of human reproduction. In 1992, at the end of August, those scientists who had negotiated the Edinburgh one-way system and the crowds from the city's annual arts festival began to assemble in the lecture theater at the Medical Research Council. This was a closed meeting for leading experts on fertility and reproduction. Sharpe was one of the delegates who had been asked to give a demonstration. He decided to air the estrogen hypothesis, partly to see what the reaction might be, but also to find out if scientists from different disciplines had anything else they could add to the idea. To his delight, the reaction was extremely favorable and there were a number of questions.

When the session was over, Richard Sharpe was preparing to go back to the lab when he was approached by a scientist he hadn't met for some years, Professor Mikko Niemi from Finland. "He took me aside and asked if I had considered changes in people's diet as a possible cause. I said no, how would diet affect estrogen exposure? He then told me about Scandinavian work which had been going on for many years. Now, I wasn't aware of this data at all. But studies in Scandinavia had shown that in women you get recycling of excreted steroids, such as estrogens from the gut. They are reabsorbed and the woman is *exposed to them for a second time*. So you literally get recycling of natural estrogens within the body."

This process of recycling and reabsorbing our natural estrogens can be influenced by diet. Herman Adlercreutz at the University of Helsinki had been studying links between diet, breast cancer and sex hormones. He was puzzled as to why hormone-dependent cancers, such as breast, colon, prostate and endometrial cancer, and some other diseases, have much higher incidence in the West than in Asia and southern and eastern Europe. It was clear from studies of migrants to Western countries that their increasing risk of breast cancer was related to adopting a Western-style diet. Most migrants rapidly abandon their traditional vegetarian or semi-vegetarian diet and consume a diet rich in calories, fat and protein, and low in unrefined carbohydrate and fiber. These and other studies suggested to Adlercreutz that the Western diet, with a high fat and protein intake and low intake of fiber and whole grain products, might actually alter hormone metabolism in our bodies. What is more, his studies suggest the changes are fundamental, altering the way estrogens are absorbed, circulated, broken down and excreted from our bodies.

To try to find out if these changes in hormone metabolism could affect cancer risk, Adlercreutz and his team had conducted an elaborate series of experiments to investigate differences in diet between women with breast cancer and healthy women. To avoid seasonal differences in dietary intake, women were studied for up to five days, four times a year. Measurements of blood and urine were taken in order to measure the levels of sex hormones. Natural estrogen in our bodies can be broken down and metabolized in the body in up to thirteen different forms, including estrone, estrone sulphate and many others. The aim was to try to build up a complete estrogen profile for each woman.[5]

He found that a typical Western diet is associated with high estrogen levels circulating in the blood and low levels of sex hormone binding globulin (SHBG), a protein which binds to the estrogens, preventing them from being active. In other words, it seems the Western diet actually *increases your exposure to your own hormones!*

"All of the changes that have occurred in our diet over the last fifty years," Sharpe concluded from the Scandinavian studies, "such as increased consumption of protein, decreased consumption of grain fiber and decreased consumption of unrefined carbohydrates, all of these changes conspire to increase this recycling of estrogens in our bodies. So this became the primary target for us. This seemed the most likely cause that we could come up with at that time for increased estrogen exposure which might explain the reproductive abnormalities."

There was a large literature going back to the early 1970s trying to establish just how estrogens are metabolized in the body. As the team in Edinburgh pored over promising charts studying urinary estrogen profiles, cross-cultural differences in estrogen metabolism, different effects of various grain fibers, it became clear there was no easy way of getting answers to their questions.

"I was highly delighted to have this idea that diet might be the root cause," Richard Sharpe recalls, "but there was one big reservation to it, which I soon realized. That is, it's untestable. How on earth could you prove it one way or another? There's virtually no way that you can actually relate diet to anything such as sperm counts, or the diet of your mother to your sperm count. So although the dietary changes look to be important, they are something which we really can't evaluate at the moment."

But as they surveyed the studies on diet, they came across another unexpected source of estrogens. "We were surprised to discover that plants make their own estrogens. They are called *phytoestrogens*." Sharpe explains, "These are really widespread in plants which we eat as part of our normal diet. Our intake of them is really quite substantial." They began to wonder whether these compounds, produced quite naturally by plants, might be involved in some way with the changes in man.

Phyto-estrogens

Investigation of estrogenic substances in plants began in Australia as early as the 1940s. In the traditionally good sheep-ranching territory near Perth, the sheep began to suffer serious reproductive problems. Worrying symptoms suddenly developed, affecting many animals in the flock. These included dystocia, a condition where the full-term fetus dies unexpectedly in the womb and the lamb is later born dead. Many ewes, even young virgin females, suffered prolapse of the womb. Some ewes failed to go into labor at the proper time, a condition which can lead to death of the lamb and the ewe. Many ewes failed to conceive at all. Curiously, young females, before they were used in the breeding program, were showing development of the udder and producing milk. Senior scientists from the government were called in to work out the cause of these widespread reproductive problems.[6]

"The economic aspects of the problem are extremely serious. Not only is there heavy annual loss from wastage of ewes and lambs," wrote one scientist in the *Australian Veterinary Journal* in 1946, "but the affected areas contain many of the best stud merino flocks in Western Australia. This will have serious effects . . . throughout the State."[7] The breeding program was halted. An investigation committee was set up and they checked out everything: the soil, possible contaminants, the number of species affected, the typical symptoms, and so on. The only clue they had to go on was that the animals appeared to be suffering a hormone imbalance. The pattern of symptoms suggested that somehow the animals were suffering from an excess of estrogens. After some time they came up with the most unexpected answer: "clover disease."[8]

The luxuriant dark green leaves of the clover seemed an unlikely culprit. However, the investigators traced the problem to a particular species of clover, *Trifolium subterraneum,* which had been recently introduced into the area from the Mediterranean. Then they isolated the active substance from the clover leaves. Laboratory tests showed it to be *trihydroxyisoflavone,* which they found could act like a weak estrogen.[9]

It was soon realized that clover was not the only plant which could produce estrogen-like substances. Professor Bradbury, from the University of Western Australia, who had been instrumental in solving the breeding problems, continued to investigate estrogens in plants. There were reports of estrogenic compounds in willow catkins, certain tubers and barks. They could be found in tulips, nettles, ryegrass and several species of clover. But what were they doing there, he wondered, and could they be used for medicinal purposes? Intrigued, Bradbury continued his investigations and by the 1950s had tracked down at least fifty species of plants containing substances that can act as estrogens, including many that we eat: broccoli, cabbage, green beans, hops, carrots, wheat, rice, vegetable oils and especially soya, which is one of the richest sources. "Our knowledge of plant estrogens is limited and fragmentary," he wrote in 1954. "Clearly a great deal more systematic investigation is required."[10]

Today many different plants have been identified which contain substances with estrogenic activity. There are several different classes of chemicals involved. The *isoflavones* are the most common. About seventy different kinds can occur naturally, though not all are estrogenic. In clover, concentrations of isoflavones can be as high as 5 percent of the leaves. The second group, the *coumestans* can be found in sunflower seeds and oils, alfalfa, soybeans, green beans, red beans, split peas, spinach, and some species of clover. These are estimated to be 30 to 50 times more potent as estrogens than the isoflavones. In fact the estrogenic activity of coumestrol is believed to be only 10 to 20 times lower than that of the natural 17β-estradiol. The third group are derived from certain *acid lactones* and are typically found in corn and wheat crops.[11]

So if phyto-estrogens could harm reproduction in animals, this raised the possibility that they do the same in humans. Once again the Edinburgh team found no simple answers to this point.

Surprisingly, there is a considerable body of evidence suggesting that plant estrogens may be beneficial in adults. Studies have shown that Asian women, who have much lower rates of breast cancer than women in the West, consume 30 to 50 times more soya products than do American women. Animal studies have also shown that rodents on a high soya diet have increased protection, when exposed to carcinogens, from breast, prostate and colon cancer compared with rodents on a diet with no soya. There are even studies suggesting it may be protective against cardiovascular disease, prostate cancer and colon cancer.[12]

It is not yet known how the phyto-estrogens may confer this protection. Some have proposed that they may act rather like the breast cancer drug Tamoxifen and block the estrogen receptor, thereby reducing exposure to our natural estrogens. Others have suggested that, unlike the high-fat diet, which lowers the production of the estrogen-binding protein SHBG in our blood, phyto-estrogens may have the reverse effect and stimulate the production by the liver of SHBG. This binds to our natural estrogen in the blood, so again it would reduce the concentration of available estrogen in our blood. It has even been suggested that they may bind to another receptor altogether, not necessarily the estrogen receptor and exert their effects by altering the biochemical activity of the cell in another way.[13]

But despite the possible beneficial effects, if these compounds could affect the reproduction of sheep, could they also account for the changes in man? This seemed unlikely. "Because we've lived with these products throughout our entire evolution we have to presume that we've adapted to their presence and can cope with ingesting them," Sharpe reasoned. After all, the changes in sperm count appeared to have occurred in the last fifty years, whereas man has been eating plant estrogens for centuries. There was, however, one caveat: "Again mankind has begun to distort the picture. Because the richest natural source of phyto-estrogens, soya, is also a very cheap form of protein, this has now been introduced on a very widespread scale into processed foods in the form of soya-derived protein."

Sharpe is particularly concerned about the increasingly widespread use of soya in baby foods. It is used as an additive to baby meals. There are also soya-based infant formulas which were developed some twenty years ago for bottle-fed infants who have allergies to cow's milk protein. Some estimates suggest that these have become so popular that as many as 25 percent of American babies are on soya-formulated baby milk. "We are now exposing infants to these in a way which hasn't happened in the past. I am seriously concerned that if soya-derived phyto-estrogens are going to have an adverse effect then the time when this is most likely to occur is in the first year of life." This is because growing evidence from animal studies shows that in the young, unlike in the adult, hormones can exert permanent programming effects.

Recent research from Auckland University in New Zealand has suggested that the biological effects of phyto-estrogens typically consumed by a baby drinking soya milk could be as much as a hundred times greater than the amount of natural estrogen the child would receive from breast milk. On a weight-for-weight basis, it has been argued, this is equivalent to giving the baby several contraceptive hormone pills a day. These and other studies have prompted further research by the New Zealand government, the World Health Organization and the Ministry of Agriculture in Britain.

Proponents of the use of soya believe that such concerns are unwarranted. They point out that Asian infants are exposed to large amounts of phyto-estrogens naturally in their diet, without apparently any adverse effects. But Sharpe remains cautious because there may be differences in Asian and Western people's responses to soya which have evolved over generations. "Because phyto-estrogens affect reproduction, you will very quickly select out, over just a few generations, people who are resistant to the effects of phyto-estrogens," he explains. "This is just natural selection. Consequently, those people traditionally eating an oriental diet may be relatively resistant to the adverse effects of phyto-estrogens, whereas we in the West are not."

In addition, it has been found that oriental babies are not necessarily even exposed to the mother's high soya diet through her milk. "One recent study on breast-feeding oriental women eating a natural soya-rich diet suggests that only a small percentage of the phyto-estrogens get into the breast milk," says Sharpe. "The amounts of phyto-estrogens the Asian infants consumed are 0.1 percent of what is present in some soya-formulated milk powders. . . . I'm not saying it necessarily has an adverse effect, but there are animal studies showing it has potential to do harm. There are also a lot of health benefits of soya, but as it contains many potent substances, its use should be treated much more cautiously than is currently the case, especially where infant foods are concerned."

But as Sharpe surveyed the literature in 1993, even with all the uncertainties over the risks and benefits of soya in both adults and infants, phyto-estrogens still seemed unlikely as a cause of the adverse reproductive changes. "These trends in food processing have taken place during the past ten to twenty years," he reasoned, "so it is hard to attribute the changes in sperm counts to them when the reproductive changes appear to have been taking place during the last fifty years." Once more, it appeared they had arrived at a dead end. Although changes in diet and phyto-estrogens may be a factor, it was hard to see either as the sole cause of the adverse changes in human reproduction.

By now, he and Niels Skakkebaek had finally published the estrogens hypothesis in the *Lancet*. It had provoked a considerable correspondence in the scientific journals. There were learned letters on how sexuality is established in the womb, on whether the same mechanisms could apply to breast cancer, and on many other related topics. There was also a spate of media interest of a rather different kind.

"It's the Incredible Shrinking Manhood!" blazed the *Scottish Daily Record*. "It's what every man fears. Top Scots boffin says manhood is shrinking." In the Edinburgh labs, these cuttings and articles would be posted on the door of the laboratory fridge or on the noticeboard. They were always good for a laugh, but beneath the bravado was considerable uncertainty. The estrogen hypothesis was no longer contained within the discrete circle of a small band of scientists. It was blazoned across the globe in language that smudged and blurred the important scientific detail. Such was the concern that Sharpe himself had even been contacted by the Department of Health for his views about risks to man. His reply was cautious but still clear: "Based on studies on animals there is growing evidence that estrogens may behave rather differently in fetal/

neonatal life and that the 'ground rules' for estrogen dose response effects derived from the adult may not be appropriate. . . . The safest attitude, as I see it, is to take every reasonable step possible to identify estrogenic contaminants in our food chain and then to minimize exposure to them. . . ."

But there were times when he was full of doubt. Had they generated a hypothesis which simply couldn't be tested? Was there no way out of the scientific impasse?

In the post room at the reproductive biology labs they were used to an ever-growing mailbag for Sharpe, with correspondence from all over the world. But it was still something of an event when a FedEx van pulled up in some urgency, with a noisy screech of brakes, in the quiet side street. There was a package for Sharpe from Washington. It was sent straight up to the lab.

Inside was a book edited by a scientist called Dr. Theo Colborn, gathering together details of talks presented at a conference of wildlife experts. The book had been sent to Richard Sharpe by a journalist in the States, Betty Hileman, who was intrigued at the astonishing overlap between the two scientists' ideas. It was to be a real turning point.

CHAPTER FIVE

Opening Pandora's Box

Washington, D.C., 1992

Dr. Theo Colborn, a scientist with the World Wildlife Fund, Washington, had edited a book which gathered together many recent wildlife studies in North America.[1] "Colborn's book was very relevant," recalls Sharpe. "I literally read the book in one night, and faxed her the next day. I think she was amazed."

"I was really surprised," says Theo Colborn. "He called me to tell me that he hadn't slept since he got the book and that he had been reading it around the clock. I was really pleased. It was exciting. I had found someone else who had understood what I had understood."[2]

"As a scientist, this is the sort of thing that you need. When you're plowing your furrow out there, your lone furrow," Sharpe explains, "you want support. It's nice to be first with an idea, but not for too long. You want confirmation that what you're working on is not unreal. Suddenly we had all this information to support us. On my way home that night I was on a real high, because suddenly our whole idea had real credibility."

The North American wildlife studies in Dr. Theo Colborn's book provided striking echoes of the changes to human reproduction which Sharpe had been investigating. It was as though they had both grasped different pieces in a complex scientific jigsaw. When the pieces were placed side by side they fitted only too well. But the picture that was beginning to emerge was vastly disturbing.

For Dr. Theo Colborn, this was the culmination of a remarkable detective trail that had begun several years earlier, when she took up a post in Washington for the Conservation Foundation. Her special assignment was to assess the environmental health of the Great Lakes. For Colborn this was the project of a lifetime. Her aim was to gather all the recent data from the Great Lakes and try to make sense of it. She wanted to know whether the health effects that could still be found in many species around the lakes were related in any way. Colborn's line of inquiry was soon to go far beyond this. She was to provide a change in perspective that would throw a sinister problem, so easily buried by the sheer wealth of disparate data, into sharp relief.[3]

The Great Lakes are the world's largest inland water mass, thousands of square miles of water, a staggering 20 percent of the world's freshwater. For mindless centuries these crystal waters, with their shimmering fish making their imperceptible journey through evolution, were undisturbed. But during the twentieth century, the regions around the lakes, with their excellent communications and rich resources, developed into the industrial and agricultural heartland of the United States. Famous towns developed along the shore side: Detroit, center of the American car industry; Chicago, with its vast complexes of steel mills and lumber mills; Cleveland; and many others. By the sixties and seventies, the Great Lakes had become infamous for their pollution. Even to the untrained eye, the devastation was obvious. The fate of Lake Erie, in particular, surrounded by chemical and manufacturing plants, became an enduring image for enthusiasts of the environmental movement. On television screens worldwide, people could see the evidence for themselves: dead fish floating in a sprawl of chemicals, detergents and oils. Birds and other wildlife washed up on the beaches of the lakes dying pitifully, shivering and wide-eyed. Rotting algae churned up on the shores with fragments of animals enmeshed in oil and debris. So acute was the problem that, on one hot day in the summer of 1969, the Cuyahoga River, which leads into Lake Erie, caught fire and burned a bridge. The river that caught fire became an international news story in the West. As if this wasn't bad enough, it was soon discovered that levels of DDT and organochlorines in Lake Michigan were even higher than in Lake Erie.

Large quantities of salmon were destroyed because of the high load of DDT they carried. People were advised not to eat fish from Lake Michigan.[4] Many regions of the lakes were pronounced officially "dead."

This was when the environmental movement was at its height, spurred on by the publication of Rachel Carson's book *Silent Spring* just seven years earlier. Carson had written: "As crude a weapon as the cave-man's club, the chemical barrage has been hurled against the fabric of life, a fabric on the one hand delicate and destructible, on the other miraculously tough and resilient and capable of striking back in unexpected ways. These extraordinary capacities of life have been ignored by the practitioners of chemical control who have . . . no humility before the vast forces with which they tamper."[5] The views expressed by Carson and others were that, somehow through our ignorance and greed, we had unleashed the vast forces of nature upon us, and we needed to clean up—fast. The burning river became a symbol of everything that was wrong. Over the course of the next few years, the use of DDT was restricted, new limits were imposed on industrial discharges into the waterways, sewage treatment works were constructed, and there were many other improvements. But the problems with wildlife were not over.

In the late eighties, Dr. Theo Colborn tracked down all the recent studies on the health of wildlife in and around the Great Lakes. The levels of industrial and agricultural waste in the lakes were much reduced, but as she pored over the data she became increasingly concerned. "Biologists working in the field were still reporting things that were far from normal: vanished mink populations; unhatched eggs; deformities such as crossed bills, missing eyes and clubbed feet in cormorants; and a puzzling indifference in usually vigilant nesting birds about their incubating eggs. . . . Everywhere there were signs that something was still seriously wrong."[6] It was a fascinating picture of an ecosystem making a recovery from the chemical onslaught of the previous decades. Clearly there had been big improvements. But, she thought, if the lakes were "clean," why were biologists still reporting so many effects in wildlife, and why were the symptoms so varied? "I felt the proclamations of recovery were premature, and came to doubt that the lakes were truly cleaned up," she reports.[7]

Intrigued, she went to the lakes and met many of the scientists involved in the research. She scrutinized the data on cancer in detail. But the studies were not clear-cut. With so many different chemicals and such apparently diverse effects in the wildlife, it seemed impossible to relate different chemicals to different effects. There was no way of even being sure that all chemicals that were out there had been identified. To complicate the picture, some of the species appeared to be particularly vulnerable to the effects of viruses, as though their immune systems had been damaged. More confusing still, when she checked the human data on cancer around the Great Lakes, there was no evidence that cancer in human populations was any higher than elsewhere. This matched the wildlife data. Overall, there was very little cancer reported in the species of animals around the Great Lakes. But the other strange abnormalities intrigued her: developmental effects, reproductive and behavioral problems, gross deformities present at birth, and defects in the immune system.

"When Rachel Carson published her book in the 1960s," Colborn says, "the concentrations of chemicals that were being used were much higher in the environment than they are today. So basically they were seeing gross mortality. Birds were dropping dead in people's yards. Carson also picked up on the problem of cancer, and that's where I think we made our mistake, because we had a knee-jerk reaction to something that was devastating, cancer, and we missed these less visible effects. I don't want to call them subtle, because they are not really subtle. They can be equally devastating."[8]

Dr. Colborn was not lightly put off. This was no ordinary academic jaded by years of teaching. She had come to science late in life. After years of raising her own family and sheep ranching, she seized her opportunity to study zoology at the age of fifty-one. One of the most mature students on campus at the University of Wisconsin, she defied the skepticism of her tutors, outpacing much younger candidates by her sheer enthusiasm and fascination with her subject. Six years later she had her Ph.D. At a stage in life when others might be considering retirement, she embarked on a career in Washington. She was used to fighting and didn't expect solutions to fall into her lap. She was prepared to work hard. Somehow, she couldn't put to rest the nagging, insistent thought that many of the symptoms reported in wildlife, although subtle and apparently very different, were in some way related. There had to be a common thread, but what was it?

Of all the volumes of data that were piling up in her office, there were a few studies which she came to ponder more and more deeply. At first glance, they were simply bizarre and made no sense. Firstly, there was a curious study which had been done more than ten years previously on female gulls. She could remember vividly her discussions with the scientists involved from the Canadian Wildlife Service.

Dr. Glen Fox had been studying herring gull colonies around Lakes Ontario and Michigan in the 1970s, when he noticed a curious change in their breeding behavior. Nests had four to six eggs, around twice the normal number. There seemed no obvious reason why this should be, until he realized that the female gulls were sharing nests together, instead of with a male, as was usual. But why, he wondered, should gulls change behavior patterns built up over years and years of evolution? Since he could find no obvious reason, Glen Fox preserved some of the chicks and eggs in his laboratory to study later.[9]

By chance, some years after this, Dr. Michael Fry, from the University of California at Davis, was interested in similar studies of female-female pairing in herring gull colonies in Southern California. Dr. Fry is a toxicologist who had been studying the effects of DDT. Between 1950 and 1970, offshore Southern California was exposed to massive contamination of DDT, with as much as 1.9 million kilograms discharged into the environment. Several studies showed high levels of DDT had accumulated in sea birds and fish in the area. He wondered, could the high levels of these contaminants account for the strange reproductive behavior of the birds? In the laboratory, he exposed gull eggs to the same levels of DDT and DDE that had been found in the wild. On hatching, although all the chicks looked perfectly normal, under the microscope there

were marked alterations in the males. The testes took on characteristics of an ovary. At even higher levels of exposure, some of the
birds were intersex, with parts of both the male and female reproductive tracts lying side by side. The males were in effect *feminized*.
In 1981, he reported his curious findings in the leading journal,
Science. DDT, he argued, was feminizing the males, reducing the
numbers available for breeding and skewing the sex ratios in the
gull colonies to such an extent that excess females were pairing
together. "Abnormal development," he warned, "induced by DDT
in birds could be more persistent than the pollutant itself."[10] It was
soon confirmed that the samples Fox had stored from the herring
gull colonies in the Great Lakes showed marked reproductive abnormalities, very similar to those that Fry had identified in the
laboratory. It appeared that the gulls in the Great Lakes, just like
the gulls in California, had been feminized due to exposure to
chemicals acting like an estrogen during development.[11]

This was one of the first wildlife reports to show graphically the
feminizing effects of some pollutants during development. For Dr.
Theo Colborn it was to provide a very important clue. There were
other studies, too, which seemed to her highly significant. She was
particularly struck by the work of Professor Frederick vom Saal
from the University of Missouri.

Vom Saal was neither a toxicologist nor a wildlife specialist, but
became drawn into the field by a chance observation. As a young
student, reading biology, one of his duties was to help with the
animal breeding program, looking after the mice. He became fascinated by the animals' behavior. For some strange reason, he
noted, there were marked differences in the levels of aggression of
the females. "A small proportion of the female mice were mounting
other females," he recalls, "attacking, biting, chasing, rattling their
tails aggressively, behavior much more typical of the male." He
found, on average, one in every six females would behave this way.
From his work in the breeding program, he knew that these mice
had been inbred for some time and had virtually the same genetic
makeup. So why, he wondered, since they have the same genes
and the same environment, can there be such a difference in their
behavior? He remembers discussing this with his post-doctoral

adviser, Frank Bronson: "Frank said, 'It's worth investigating, but you're probably not going to find anything!' Well I expected not to find anything either." But they did.[12]

The strange puzzle of the aggressive females determined the direction of vom Saal's research for many years. The answer was stranger still. Through an elegant series of studies, he showed the key to the behavior of the mice depended on their *position in the womb*. In an average mouse litter of twelve pups, roughly one in six of the females will be placed between two males. This position, it turned out, was absolutely critical. During their development, the male pups' testes secrete the hormone testosterone, which triggers masculinization of the reproductive tract. However, some of this testosterone will wash over the adjacent female pups, influencing their development too. Since testosterone not only influences the development of the male reproductive tract but also masculinization of the brain to produce typically male behavior, this, he realized, could account for the behavior of the aggressive female mice. Males positioned between two females were equally changed, with alterations in the prostate, seminal vesicles, scent-producing glands and hypothalamus: "A whole suite of differences," he explains, "in brain structure, reproductive organ structure, sensitivity of cells in their body to hormonal stimulation, behavior differences, enzyme differences, the whole way these animals would handle hormones in the future was programmed before birth."[13]

The most remarkable aspect of the study was the levels of the hormones that produced these effects. This could be shown by taking blood samples from the fetuses. "A male positioned between two females received just a twenty parts per trillion increase in estradiol in the womb compared with those positioned elsewhere. Now remember less than one hundredth of this is biologically active. So we are talking of a difference between two males of *less than one ten-trillionth of a gram*! . . . A female positioned between two males got a billionth of a gram of extra testosterone compared to her sisters."[14]

He went back to his former adviser and showed him the data. "But the differences are *so small*," said Frank Bronson.

Vom Saal looked at him. "Yes, Frank, but the consequences are *so dramatic.*"

"I can't argue with that."

"These studies demonstrate that hormones operate during development on the fetus at levels of sensitivity that can scarcely be imagined," explains vom Saal. "These tiny differences were sufficient to alter the activity of the genes in development and these are permanent and forever consequences. Hormones change the chemical composition of the genes, quietening some, turning others on and also setting the rate of activity of the gene, for life."

Theo Colborn, reviewing this years later, was equally struck by the result. "I realized how it just takes the slightest shift in the ratio of estrogens and testosterone in the womb to change the course of development," she says. "I could begin to see at how many points during development change could take place that would start a whole cascade of events that would change development permanently. I began to realize that these chemicals are only in our bodies in low concentrations, but these are concentrations that could make significant differences while the embryo is developing. They are irreversible, and they are delayed and so may not be recognized until an individual reaches puberty or adulthood."[15]

Theo Colborn, who was not an expert on hormones during development, started to read books on endocrinology. She was beginning to suspect that, to understand what might be happening in the Great Lakes, she needed to know more about hormones. She reached a point where she had read well over two thousand scientific studies. There were extraordinary parallels between the different studies, she recalls. She was certain there must be a connection.

"There were at least sixteen top predator species in the Great Lakes that were showing reproductive problems: population decline, fertility problems, immune problems, all developmental problems," she explains. "You could see it in the offspring: young chicks born with adult plumage. Something went wrong during that embryonic development. If you start looking around the country, you'll find populations of sturgeons that are just old sturgeons: they're not replacing themselves, they're not reproducing. Go to

Florida and look at the Florida panther, many of them have developmental problems, undescended testes and so on. Fish in the Great Lakes have thyroid problems. As a matter of fact, in Lake Erie today, the thyroids are getting so large they are exploding. Also the male fish do not reach sexual maturity in the Great Lakes; the fish seem to be hermaphroditic. If you start looking, you can find fish with both male and female gonads in the same species. Below paper and pulp mills the males are being demasculinized and feminized and the female fish are also being masculinized and defeminized."[16]

But why such diverse effects? It was very frustrating. It still didn't make sense. Her office became a sprawl of scientific papers and notes. She put all the data on the computer to try to see a pattern between species that had the greatest range of problems. "I put it into a huge matrix to make some sense of it," recalls Theo Colborn. "And as I began looking at it, it was very obvious that it was the developmental effects, those effects that prevented the animals from reaching maturity, or premature death, that was causing the problem. *It was in the offspring.* The hormone system suddenly struck me as being probably one of the underlying physiological systems that was being affected. It was overwhelming. Suddenly I looked at this and said: 'Oh my goodness, what have I found?' "[17]

This sudden recognition was so strong and so powerful that she was quite shocked. As she studied the matrix, it was very clear that many of the symptoms still being found in wildlife could be explained by hormone disruption. Somehow, she thought, chemicals were affecting development by altering hormones. *Hormone disruption was disturbing development!* The picture was so stark and clear that she wondered why she had never seen the connection before. It made sense of all the studies she had puzzled over during many months.

"As I began to see the underlying problem here, I was certain that no one would believe me," Colborn recalls. "It was one of those things that was just too bizarre."[18] She began discussing her ideas with endocrinologists and searching for other examples from wildlife and human studies. After a while, she became so concerned at the evidence that she took an unusual step. She decided to organize a scientific conference to bring together all the leading experts from the different fields.

Frederick vom Saal, whose studies on hormones in the womb had had such an impact on her, recalls: "I got a phone call and it was Theo Colborn, who of course I'd never met. She said, 'I've read a paper by you, and I think that what you're doing provides an answer to what is happening to wildlife in the Great Lakes that are showing abnormal reproduction. Have you ever thought about this?' " Vom Saal explained that he had not. "So Theo Colborn presented a whole bunch of information to me. And looking at the data, I was astonished at what she had found in wildlife. I spent a few weeks looking at it and then called her up and said: 'My God, this is really something!' "[19]

Shortly after this, she was invited to a scientific meeting in Raleigh, North Carolina. Professor John McLachlan, who was the first to raise concerns about environmental estrogens, was invited to the same meeting and they were introduced. It was soon very clear that they had a lot to talk about. "I, quite frankly, was not aware of all the things that had been observed in wildlife," he recalls, "and she, quite frankly, was not that aware of the effects of estrogens on laboratory animals and humans and the DES model." They soon left the meeting to discuss ideas. "She seemed very knowledgeable about her part of the field, very serious, very intense. She was very savvy about combining ideas," recalls John McLachlan. "She could see the overlaps between the laboratory studies on estrogens and sexual development and the wildlife biology. That was a very important first step." They sat outside and discussed plans for Colborn's conference.[20]

The meeting took nearly two years to organize. Eventually she invited over twenty specialists to Racine, Wisconsin, on 26 July 1991. In the run up to the scientific conference she distributed to each participant details of the publications and background of all those who would attend.

Dr. William Davis, a research ecologist with the Environmental Protection Agency, remembers calling her before the event. "Theo I'm really honored to be invited to this, but I don't fit," he said.

"What do you mean?" she asked.

"Well you've got really renowned and senior endocrinologists and biochemists here. Let's get it straight. I'm a fish watcher. I go out and study fish. Their behavior, toxicology and physiology. I'm not anywhere near the sort of people who can define this on a chemical basis."

"And that's why you're coming," she insisted firmly. "You're coming to help give the wildlife perspective."[21]

It was, in fact, the first time that so many different specialists in this area had been brought together: wildlife biologists, endocrinologists, molecular biologists, biochemists, neurologists, marine biologists. Over the course of three days, each presented their evidence to the others in turn. Many of them had been tackling different fragments of the puzzle and were unaware of each other's work. Undaunted, Dr. Theo Colborn had asked them to present any findings that were relevant to the idea of *chemically induced alterations in sexual development*. It turned out to be, for many of them, a very special event.

"I remember sitting at the meeting, which was at Frank Lloyd Wright's last prairie home that he designed in the 1930s for some industrial magnate. These prairie homes were like ships, so that if viewed from some vantage point, they looked like some huge ship plowing through a field of wheat. That was the first thing that struck me," McLachlan remembers. "But what really struck me the most was sitting there in the audience hearing discussion of feminization of seals in the North Sea, or gulls on the west coast of America, or fish in the Great Lakes and being just *stunned* by the common biology we were seeing . . ."[22]

There were many distinguished speakers, such as Professor Howard Bern, from the University of California, Berkeley, a very senior figure in the field of endocrinology.* He had been one of the first to warn of the potential dangers of the synthetic estrogen DES, which had been given to pregnant mothers in the 1950s. There was

---

*This was a closed meeting. Rather than an exact record of the science discussed, the following description is drawn from the record and takes account of subsequent papers and interviews to give a full account of the wildlife evidence.

so much to be learned from the tragedy, he argued. Firstly, exposure to a synthetic estrogen early in fetal life can create abnormalities in the development of the reproductive tract, such as the vaginal cancers seen in young women. Secondly, there are critical periods where exposure to estrogens can be much more disastrous. Human vaginal clear-cell cancer, for example, was more likely to arise if exposure to DES was during the first trimester. Most worrying of all was the subtlety of these effects. No one looking at a DES-exposed baby at birth would be able to detect any "birth defect." The changes that lead to the cancer do not appear until adolescence or even later, showing that the effects of synthetic estrogen exposure can be masked for some time.[23]

Like John McLachlan, his studies on DES had led him to worry about environmental estrogens. "There is the possibility of exposure [to estrogens] from therapeutic and contraception procedures, from ingestion of foodstuffs, and from contact in the environment with substances that are estrogenic," he warned. "Such agents may have developmental effects that will parallel the DES syndrome." He urged for more studies to protect "the fragile fetus."[24]

John McLachlan, too, discussed environmental estrogens. He presented his latest ideas on how synthetic estrogens might interact with the estrogen receptors. It was a subject he had discussed publicly many times before. But the context today gave it much greater impact. He began by explaining our increasingly complex chemical environment. "While our external chemical·milieu has changed dramatically in the last half century, many millions of years were required for living organisms to evolve strategies to deal with their internal environment. There are many ways to interpret environmental signals: one is through the hormone receptors. . . ."[25]

He went on to outline his thinking on how the hormone receptors work. The receptors, he explained, appear to be vehicles for turning on and off a cascade of events as a result of exposure to estrogenic chemicals. He described the chemicals that appear to have estrogenic action and pointed out the puzzle that they have different shapes. He showed why he believed that exposure to estrogens could not only bring about obvious physical abnormalities, but could also "feminize the animal at the molecular level," and outlined his work on lactoferrin. With characteristic understatement, he concluded: "Chemicals in the environment with estrogenic activity may have long-lasting effects on sexual development, both prominent and subtle. . . ."[26]

Professor Frederick vom Saal also gave a talk outlining his remarkable studies showing that tiny differences in hormone levels arising from a fetus's position in the womb could have effects on development. "People think of hormones as activating a transient event, such as cueing the release of the egg from the ovary. But in the fetus, hormones don't have a transient effect: *they program the way genes function for the rest of that animal's life*," he explained. As he talked about the tiny concentrations of hormones required, parts per trillion and lower, it was easy to envisage that chemicals in the environment, even though present at low levels, could be cause for concern. "Estrogenic compounds," he argued, "can have devastating effects on development, although these effects may not become apparent until the individuals reach adulthood or even old age."[27]

"I was really struck by Frederick vom Saal's talk," recalls Dr. William Davis. "It was fascinating. I was very taken with the very low levels of estrogens and testosterone in the womb that could make a difference to the eventual behavior and life history of the animal. I hadn't any idea that someone was doing that kind of work."

The conference was beginning to take on a momentum of its own. It was now becoming very clear that many scientists studying the effects of hormones and hormone mimics in the laboratory had been thinking on exactly the same lines. Even more striking was the amount of data on wildlife which appeared to show that the effects identified in the laboratory were really happening in the wild. Every report seemed to highlight ways in which that ancient and exquisite balance of nature was disturbed. The whole world, it seemed, awaited the threatening touch of something malevolent and unknown.

Dr. William Davis and his team from the Environmental Research Laboratory in Florida had gathered evidence of *masculinization* of certain species of fish. Researchers were first alerted to this in 1980 when studying Eleven Mile Creek, in Florida. Near the head of this creek was a paper and pulp mill which discharged thousands and thousands of gallons of effluent on a daily basis. Downstream of the discharge, researchers were startled to find pregnant female mosquito fish with typically male characteristics. The most obvious changes were the greatly enlarged anal fins of the females, which had developed into a structure called a "gonopodium," normally only found in males. This appendage serves to transfer and insert sperm for internal fertilization. These "highly masculinized" females showed male fish behavior, including mating attempts. There were other details of masculinization: the telltale alterations to pigments, spots, markings and fins. Further studies by the Florida team had confirmed that, far from being a scientific curiosity, such alterations in the fish could be found in many streams into which pulp mill effluent was discharged.

"Masculinization . . . appears to be permanent," they observed. "Exposure during the early stages of the life history appears to produce the most severe responses, including potentially hermaphroditism." The cause, which substances in the mill effluent were doing this, has not been determined. They also suspected other cases of "pollution-induced sex alteration," such as the early maturation of the American eel. There was some evidence that eels were being stimulated to mature years early. Eels not much bigger than earthworms were showing precocious maturity, with secondary male sex characteristics: the enlarged eye and changed pigments characteristic of larger eels preparing to migrate later in the life history. If, he speculated, they were prompted by this precocious maturity to begin their migration to the Sargasso Sea in the middle of the Atlantic, they would not have the body reserves to survive. Their studies highlighted the possibility that environmental hormone mimics could indeed disrupt reproduction in wildlife.[28]

Further impressive wildlife evidence came from Peter Reijnders from the Research Institute for Nature Management, Den Berg, Holland. He, too, made the marine environment sound like a reproductive battleground. He had reviewed many studies identifying

changes in levels of hormones, such as progesterone and testosterone in starfish, reduced levels of estrogens in English sole, reduced
levels of testosterone in Atlantic cod and winter flounder, impaired
ovarian growth in the Atlantic croaker, decreases in testosterone in
porpoises, and changes to mussels, plaice and whelks. Higher up
the food chain, there were changes in hormone levels in marine
mammals: Baltic ringed seals, gray seals, harp seals, premature
pupping in California sea lions, and reproductive disorders and hormone disruption in the St. Lawrence beluga whale. He had studied
sudden die-offs of thousands of dolphins, porpoises and seals, particularly in the Baltic and North Seas in the late 1980s, and examined the evidence that immune and reproductive failure may be
due to chemical contamination. Hormonal imbalances are observed
throughout the entire marine ecosystem, he explained, although it
is proving difficult to link any observed effect with any one given
pollutant. The effects of the hormonal imbalances could be seen in
a suite of disturbances to reproduction: lowered egg production,
decreased ovarian growth, hermaphroditism, implantation failure
and disorders of the genital tract from cod to croaker. The litany
of strange disorders and subtle deviations from the norm was reminiscent of the kind of weird creatures that inhabit a Lewis Carroll
novel.[29]

Although in many cases the contaminant could not be proven,
Reijnders was particularly concerned about the continuing effects
of PCBs and DDT. His earlier studies had attributed reproductive
failure in common seals to PCBs, showing the most significant
damage to reproduction was in seals inhabiting the most contaminated areas with high levels of PCBs in their tissues. To check this
further, he and his team fed two groups of common seals different
diets. One group was given plaice and flounder caught in the more
contaminated Wadden Sea. The second group ate much cleaner
fish from the northeast Atlantic. Tests on the diets confirmed that
the fish from the Wadden Sea contained on average seven times
more PCBs than the fish from the Atlantic. After nearly two years
on these diets, there were striking results. The group of seals fed
fish from the contaminated waters produced much fewer offspring.
Ten out of the twelve females eating Atlantic fish became pregnant

and produced seal pups. Only four out of ten females in the group eating the more highly contaminated herring became pregnant. Further studies revealed that the damage was occurring around the implantation stage.[30]

"It's estimated that 1.5 million tonnes of PCBs have been produced on a global basis since the 1930s," he warned. "Calculations suggest that 20 to 30 percent of all produced PCBs . . . have found their way into the environment . . . Even though global production of PCBs has now stopped, more than 70 percent of global production is still in use and could reach the environment in the future . . . Unless adequate retrieval and disposal of remaining PCBs in use is enforced, these noxious compounds will become an even larger environmental problem than at present."[31]

Further concerns were expressed by Glen Fox, from the National Wildlife Research Center, Hull, Quebec, whose research on the "feminized" herring gulls had helped Theo Colborn shape her ideas. These classic studies, finding female gulls sharing nests, had been the first clue in a detective trail that led eventually to pollutants like DDT. He, too, shared the concerns about the studies on fish near paper and pulp mills. Researchers had noted masculinized female fish could be found as far as four miles downstream. There was also some evidence of changes to birds feeding off these fish in some areas near the pulp mills, with eggs failing to hatch and offspring not being raised. He was also intrigued by another curious study which highlighted sex alteration in wildlife.

A strange syndrome had been identified in marine gastropods, marine snails and dog whelks, by a biologist, B. S. Smith, and his team ten years previously. Studying the humble mud snail, they had observed the females had male secondary sex characteristics literally imposed on their own sex organs. Female snails were identified with penises and a penal duct, a vas deferens, and changes to the oviducts. Smith coined the term "imposex" to describe this bizarre syndrome. This causes sterility and sometimes death. It was soon realized that this condition was very widespread and could be found in up to forty-five different species of snails, dog whelks and other gastropods all over the world, from the Plymouth Sound in England to the distant shores of Alaska and the Far East. Research-

106 / Altering Eden

ers eventually tracked down the problem to biocidal paints used on boats, containing the compound tributyltin. Tank experiments on these paints showed that the chemical used could exert hormonal effects at exceptionally low concentrations and this had led to a partial ban on these anti-fouling paints in some countries. But this study highlighted, once again, the very real possibility that chemicals in the environment could affect sexuality.[32]

Fox concluded that the female-female pairing in herring gulls, the masculinization of fish near pulp mills, and the development of a penis in female snails provide "conspicuous" flags signaling otherwise subtle alterations in sexual development. These changes have severely affected the reproduction of some populations and can be geographically widespread. "It is time to ask, *what effects have these and other chemicals had on human populations? And more importantly, what can be done to detect, control and eliminate environmental chemicals capable of altering sexual development?*"[33]

The talks continued, each adding more to the wealth of data. There were other studies, too, to consider, from those who had not been able to attend. John Leatherland, Chairman of Biomedical Sciences at the University of Guelph, Canada, had been investigating thyroid changes in salmon in the Great Lakes. Among his most disturbing findings were the grossly enlarged thyroid glands in salmon from Lake Erie. The thyroid gland, which regulates metabolism, can normally be found in fish as little follicles scattered near the lower jaw. In the Great Lakes salmon he found the thyroid glands could stretch from the tip of the jaw to the tip of the heart, bulging out of the gill arches and sometimes invading major blood vessels. "They are really huge, the size of a golf ball," he reports. At first he suspected this was an iodide deficiency problem, but he found the fish in the Great Lakes had perfectly acceptable levels of iodide available to them. It became clear that birds eating these fish can also develop thyroid problems. "Clearly there is something there," he says. "I think an environmental cause is the most likely."[34]

Taken together, the wildlife evidence highlighting hormonal alterations in many different species had a profound effect, especially for scientists like Professor John McLachlan who had been con-

cerned about environmental estrogens for many years. "I was literally astounded to hear the case reports that were coming in from
colleagues working in the field," he recalls. "The wildlife biologists
were describing the feminization of a variety of different species
which were linked to environmental estrogens. The studies with
birds and fish were remarkable. . . . I never really interacted with
the wildlife community and yet they were describing in wildlife
exactly what we could show in the laboratory. Some of the studies
were compelling."

"Unlike the rest of the scientists who came in absolutely stone
cold in terms of what was likely to happen, I went in having talked
to Theo on almost a daily basis for six months," says Professor
Frederick vom Saal. "I still came away from the meeting absolutely
stunned by what was very clearly a very serious problem. Theo
came away stunned. None of us was really prepared for the way
the information synthesized. . . . Let anyone argue with me and say
these wildlife studies aren't relevant to humans."[35]

There were also compelling similarities between different studies.
For instance, Richard Peterson and his team from the University
of Wisconsin had undertaken studies showing how dioxin exposure
dramatically affected reproductive development. Like vom Saal's
twin studies, this showed the extraordinary sensitivity of the developing fetus to chemical disturbances. Neither vom Saal nor Peterson had come across each other's work beforehand. Yet they
found at the meeting they had done studies which had led to very
similar conclusions. It was just one of many successful collaborations
that were to emerge from the meeting.

The sessions continued, with Professor Ana Soto from Tufts
University, Boston, Dr. Michael Fry from the University of California at Davis, Dr. Melissa Hines, Dr. Phyllis Blair and Dr. Richard
Green, each raising different aspects of the same theme. Dr. Klaus
Dohler from Hanover, Germany, discussed the influence of hormones on sexual differentiation of the brain and central nervous
system. Dr. Melissa Hines, from UCLA, was studying the effects of
estrogens on neurobehavioral development. Dr. Phyllis Blair, from
the University of California at Berkeley, presented her findings on
how exposure to DES can affect the immune system and may

increase susceptibility to auto-immune diseases.

"We began to see that other people were seeing the things that we saw. The shadows, the images . . . they began adding up as being reality, not just glimpses," says Dr. William Davis. "We got whole new pictures, avenues and windows to look out of. . . . It got so fascinating because we began to see the consequences of all this. . . . It was like taking one step over the cliff. You knew how far it was down below, but you didn't know whether you could fly or not."

For Theo Colborn, it was an extraordinary few days. Only ten years ago she had been studying for her degree and regarded with some skepticism as a mature student. Yet here she was, with some of the world's top experts on environmental endocrinology, who were congratulating her for recognizing the parallels between the different areas of research and drawing the teams together.

Toward the end of the conference, she arranged for the participants to be split into groups to write reports. Recalls McLachlan, "I was the chairperson of our group, and Michael Fry was the rapporteur who was to report our group's findings back to the conference. We were stuck in this room. I was feeling claustrophobic. It was a beautiful day. We could see the woodland outside. We were close to Lake Michigan. So I schemed with everyone that we would take bikes from the conference center and take off to Lake Michigan. When we sneaked out and jumped on the bikes, officials from the conference center came running out, chasing us and shouting, 'Come back! Come back here! You can't do this!' But we were off, zooming away on our bikes. It was almost like in a movie."

As they went along to the lake, they started to compose crazy verses of rap about environmental estrogens. Soon everyone caught on to the idea. It wasn't long before they had created a rap song.

"On the last day, we had to present our report," says McLachlan with a smile. "People were sitting around in this very elegant conference center. Everyone was very serious. We were the first group and were asked to give our report. So I said that since we had a rapporteur we decided to give our report in rap. Michael Fry, who must be fifty something, got up, pulled a baseball cap out of his back pocket, put it on his head backwards and delivered the rap song."

At first everybody was totally stunned and disbelieving. Such exuberance was not the usual hallmark of a scientific conference: gatherings better known for their white-coated conversations and lecture-room dedication. But this *was* different. The sessions revealing the common biology had been so extraordinary. It was impossible to suppress high spirits. "Suddenly they all went completely crazy," says McLachlan, "and started to laugh and applaud. We handed in our written report, which I'm sure was as brilliant and insightful as usual, and that saved us."[36]

Each group presented their reports. With the extraordinary momentum of the last three days, everyone agreed they faced an overwhelming problem. "Basically," concluded Theo Colborn, after listening to all the evidence, "we are talking about a matter of survival, not only wildlife, but humans as well. You can reach a point of no return where it is too late and there is nothing you can do. We're going to have to decide how long we want to wait, and how much more evidence we want to collect, before we do something. This could determine how successful we will be in turning things around."[37]

A few months later, a written consensus statement was agreed upon by all the scientists. Their conclusions were stark and unequivocal.

> We are certain of the following. A large number of man-made chemicals that have been released into the environment, as well as a few natural ones, have the potential to disrupt the endocrine system of animals, including humans. Among these are the persistent compounds that include some pesticides, industrial chemicals and other synthetic compounds. Many wildlife populations are already affected by these compounds.[38]

Back in Washington, a year after this, a book of the conference proceedings was prepared and ready for release. Several of the scientists at the conference presented their conclusions to a meeting of experts from federal agencies, universities and international commissions dealing with human health. An article about their concerns appeared in the *New York Times,* quoting McLachlan, Fry and Colborn.

But by a curious coincidence, that very same week, a major five-part series on pollution was also run in the *New York Times,* asking: "What Price the Clean Up?" Beginning under the headline, "New View Calls Environmental Policy Misguided," the series illustrated the claims of scientists and health experts who argued that billions of dollars are wasted each year in battling with environmental problems that are not especially dangerous. This policy, it was argued, evolved largely in response to popular panics. The cost of the environmental program of about 140 billion dollars a year, 100 billion of this from industry, may not be justified, claimed the experts cited. Are we getting value for our money? This series, questioning the need for such an expensive environmental program began on the front page and was very prominent in the paper.[39]

Was it just coincidence, Colborn wondered, that such a series should come out the very week she published her data showing the need for more careful testing of chemicals, looking for long-term and developmental effects, not just cancer? She began to feel uneasy. As scientists, they had written exactly what they had found and believed to be true. "Would anybody try to undermine it?" she wondered.

In Edinburgh, a copy of her conference proceedings arrived at the Medical Research Council. Closeted in the tiny MRC library, surrounded by stacks and stacks of learned medical journals, Dr. Richard Sharpe was totally absorbed in the details of the conference. The immense complexity of the problem was beginning to take shape. Firstly, Sharpe and Skakkebaek in Europe were unraveling a picture of the human data: the evidence suggesting that sperm counts were declining and disorders of the human reproductive tract were increasing. Secondly, they now had North American wildlife data which mirrored many of the findings in humans: birds, fish, panthers, seals, all with reproductive and developmental problems. What is more, the scientists researching this data were even proposing the same hormonal mechanism disrupting development! Thirdly, there was all the laboratory data: the animal studies which showed graphically how these effects can be induced by exposure to tiny amounts of estrogenic chemicals. The weight of evidence was beginning to add up.

But in spite of all of this wealth of animal and laboratory data, there still was no *proof* that chemicals were harming man, he thought. Just as with his suspicions about diet and phyto-estrogens, there was no obvious way of testing these ideas. He discussed the data with his team and no one could see a way out of the impasse. Which, if any, of the chemicals to which we might be exposed was causing the effects? The complexity of the problem was overwhelming. There were so many different chemicals that have become woven into the fabric of our lives in the last fifty years. "How could we test any of this in man?" he wondered. "How on earth could you work out the exposure of a person's mother to PCBs, for instance, or even the exposure during childhood. How could we test what the effects of parental exposure to any of these chemicals were in the offspring? This was the big problem. There actually was no straightforward way to address that question. It was impossible to prove which of all these chemicals, if any, was affecting man's reproduction. So, again, we seemed to be at something of a dead end. . . ."[40]

But as luck would have it, in England, secret research was under way, research that had been known about all this time in Whitehall. This provided a rather unexpected way forward.

CHAPTER SIX

Secret British Experiments

London, England, February 1993

Professor John Sumpter had finished his teaching for the day. He left the biochemistry lab, with its neat rows of brown benches, Bunsen burners and taps, in the hands of the lab staff, and walked down the gleaming gray corridors of Brunel University to his office. On the noticeboard were brightly colored photographs of wildlife, especially fish, the brilliant splashes of reds, yellows and silvers seeming all the more prominent against the institutional background. John Sumpter is an expert in marine biology.

"If you ask me why I got interested in fish," he explains with a smile, "I can't answer that question. I do remember as a child being given a fishing rod when I was very young. It came from Woolworths and cost five shillings and I can still picture it now." They lived on the seashore in Portsmouth and his youth was spent fishing. It was the beginning of a love of natural history and a passion for nature that would dominate his life. After collecting a Ph.D. he moved to Brunel University and his career progressed rapidly. He found the intellectual challenge of research totally absorbing. Who, after all, could anticipate that unexpected changes in the reproductive life cycle of the trout would turn out to have extraordinary relevance to man? But this was the issue that was on his mind as he planned his call to Dr. Richard Sharpe.

John Sumpter was planning to take a step which was a little unusual in science. As scientific knowledge has progressed in the twentieth century, each main subject has blossomed into literally tens of different fields. Not surprisingly, it can be very difficult for a scientist to keep up with all the developments in his subject. In some cases, specialism is so great that even the language and concepts may be unfamiliar to people outside the specific field. Consequently there exists the curious state of affairs whereby the more knowledge is accumulated and the more vast the database, the easier it is for a scientist to be ignorant of developments which may be extraordinarily relevant. Not so, John Sumpter. He knew the literature backwards.

He was only too aware of the data on declining sperm counts. He had read and reread Sharpe's latest papers and felt that the discoveries that his team had made could be highly relevant to Sharpe's work. Despite this, as he prepared to call Richard Sharpe, he felt apprehensive. He had admired Richard Sharpe's work for several years. "He has an extraordinary skill," says Sumpter, "just in doing the right thing at the right time, before other people have recognized its scientific importance. He's extraordinarily successful at it." Normally, ideas would be exchanged at conferences or in letters to learned journals. It was a little unusual to call someone you didn't know, working in a different field, quite out of the blue. Besides, until recently it would have been unthinkable, because his research was still classified. He filtered some water—he more than almost anyone in the country was acutely aware of its contents—and made a cup of strong coffee.

"I plucked up the courage to ring up Richard Sharpe," recalls Sumpter, "who is probably the country's leading authority on male fertility and put the proposition to him that there might be a connection between estrogens in the environment and male fertility. . . ."

"It was astonishing because I'd been thinking about nothing else for about the past year," says Richard Sharpe. He had not yet published his ideas on exposure to estrogens as a possible cause of human reproductive problems. Yet here was a complete stranger working in a completely different field, coming up with the same theory.

"It was clear within a few seconds of speaking to him that we were both very interested in what the other one knew and that we both had very different aspects of what might be a similar story," says Sumpter. "I remember going quiet on the phone and both of us realizing that we perhaps had more than we ever expected. . . ."[1]

The story that John Sumpter had to tell Richard Sharpe was all the more exciting because it appeared to provide a very elegant means of testing which chemicals might be relevant to humans. This could be the breakthrough they needed.

It had all begun several years ago. In field studies of the River Lea in North London, they stumbled on a curious finding. Male fish were showing odd characteristics, more like a female. When they took blood samples from these fish and analyzed them in the laboratories, the results were astonishing. The males were making very high levels of the female egg yolk protein, vitellogenin. This is a white milky substance, which in females is used in the ovaries to build up eggs to make yolk. Since normal males don't make this protein at all, they suspected something was up. But was it their technology at fault, or were the fish really being *feminized*?

"You can't do science without having some idea of what results you expect. So when you get a result you don't expect, your first reaction is 'What's gone wrong?' " says John Sumpter. At this stage, no one in the world had published anything about "feminized" fish. Sumpter's team had just developed new sophisticated techniques for testing for vitellogenin in fish. They spent about two years checking every aspect of their work, assuming the result must be a technical artifact.

But each time they tested male fish from the River Lea, they got the same result. "The levels of vitellogenin in the blood of the males were extraordinary high, increased by about 100,000 times. The levels reached those we would normally expect to find in a fully mature female trout which was making lots of eggs. More than half of their blood protein was yolk. It was extraordinary. The male animal was acting very much like a female. Usually the eggs are up to 25 percent of the body weight of the trout, and so the whole physiology of the fish had been changed from being a male which would normally produce sperm to one that thinks it's got an ovary and must produce yolk." The males in effect were *changing sex*.

Stranger still, they found the effects were most marked near sewage outfalls. It was beginning to look as if something in *human* sewage was capable of "feminizing" fish. After about two years they could find nothing wrong with their methods and were beginning to conclude, however bizarre, the result must be correct.

Through a chance conversation, they learned that they were not the only ones who were aware of this problem. British government officials had already been notified. In the late 1970s anglers had reported the presence of hermaphrodite fish to the Department of the Environment. These are fish which are intersex. Instead of having either ovaries or testes, the gonads were a mixture of the two. Normally such fish are rare, but there were several reports of anglers catching these fish in the lagoons where effluent from sewage treatment works flows before it enters the river. It was unclear at the time whether these effects were natural or arose because of something in the effluent from sewage. Soon after this, scientists from the Lowestoft laboratories of the Ministry of Agriculture had been surprised to find greatly increased levels of the female egg yolk protein, vitellogenin, in male fish in one of their farms located just downstream from sewage treatment works. It was beginning to look as though, far from being wrong, their technology was indeed a sensitive test of what could indeed be a wider problem.

The Lowestoft and Brunel teams approached the Department of Environment for funding to test this further. Soon it was official and a joint project was set up. Between them, they designed a unique experiment. Their aim was simple: to find out if sewage effluent contained sex hormones that could be feminizing fish. Executing the plan was much more complex.

An English meadow in the summer may appear full of the enchantment and promise that prompted Wordsworth, Shelley and Keats. But in reality, for the biologists armed with buckets of fish and cages, there were a number of undesirable extras to be negotiated: knee-high nettles, brambles, bogs, mosquitoes, angry owners, dogs and "keep out" signs. Even when these obstacles had been overcome, they had to search for a detail the poets overlooked: the point on the riverbank where sewage effluent was discharged directly into the river. Although this was treated effluent, the smell was overpowering. Protected by rubber waders they would scramble down the bank into the fast-flowing water and attach the cage in the river water close to the outflow. The rainbow trout were then transferred, about twenty in each large case. The conditions made this definitely not a job for anything other than an enthusiast!

But their efforts were well rewarded. After just three weeks' exposure to human sewage effluent, the fish were removed and blood samples were taken back to the laboratory for analysis. The results confirmed all earlier suspicions. The males were making very high levels of the female egg yolk protein, vitellogenin. Since vitellogenin can only be made in response to the sex hormone estrogen, the results strongly suggested that the male fish were being exposed to something that could *act* like an estrogen.

"Male fish do have the biochemical machinery for making yolk protein inside the liver," explains Dr. Charles Tyler from Brunel University. "But normally, the machinery lies idle, simply because there is no estrogen to switch it on. But if they are being exposed to estrogens some other way, the machinery switches on and yolk starts to build up in their blood."[2]

"For me the results were quite astonishing. I had never even heard of environmental estrogens before this point," recalls Dr. Peter Matthiessen, Head of the Biological Effects Group at MAFF fisheries laboratories. "Yet, the levels of yolk protein in the male blood were as high as levels found in gravid female fish that are producing eggs for spawning. This represents a huge metabolic cost to the male fish. All the energy of the male animal has been subverted into producing yolk protein instead of putting it into more normal male pursuits such as sperm production and chasing females."[3]

"Before long," remembers Dr. Tyler, "we had a major detective story on our hands, with a plot and named cast of suspects worthy of any conventional whodunit. . . ."

They double-checked these findings on rainbow trout with another fish, the carp. The results were exactly the same. So, they embarked on a nationwide survey. They began discussions with all ten water authorities to find three or four suitable sites in each region. During the long hot summer of 1987, the effluent from thirty-one different sewage treatment works in both England and Wales was studied.

"We got very positive results at all of the sites that we tested," explains John Sumpter. "So the effect is not one localized to just one particular area of the country. It's a nationwide effect and a very, very pronounced effect. To me this was the biggest surprise, because nobody imagined that it would be a nationwide problem. *The only conclusion from the finding must be that there is something in the effluent coming from sewage treatment works that acts like the female hormone on fish. That's the only explanation."*

What is more, the substance appeared to be so potent that in some rivers it was having effects a considerable way down from the discharge sites. "We've done a series of surveys on the River Lea where there are a number of sewage discharges," explained Peter Matthiessen. "We've found that there are elevations of vitellogenin in caged male trout all the way down the river, with particular peaks just immediately downstream of the various sewage discharges."

They had other worries too. "Clearly any substance which is diverting the energies of an animal into producing large quantities of a protein that it doesn't normally make must be causing other knock-on effects in that animal, not least the sheer metabolic cost of it," says Peter Matthiessen. "So we thought there could well be other effects, including reproductive effects. *What we also wanted to know was, are these substances getting through from rivers into our drinking water?"*

In the world of science, it is widely accepted that any scientist with a particularly interesting result aspires to publish it in the journal *Nature.* Here, under the knowing but withering gaze of John Maddox, who recently retired, standards of scientific excellence have been set for years. A paper in *Nature* has always been a hallmark of distinction. "Colin Purdom, one of the senior scientists on our project at that stage, Deputy Director of the MAFF labs in Lowestoft, was due for retirement," recalls John Sumpter. "He wanted to go out on a paper in *Nature.* He wanted to go out on a high." But, unfortunately, this wasn't possible. When they informed the officials at the Department of the Environment about their publishing plans, they were in for a surprise. Their research was *classified*.

"We found out that the project from its inception at the beginning of 1986 had been deemed confidential," says Sumpter. "This was at the insistence of the water companies. So it wasn't possible to publish."

Several years had now elapsed since Sumpter's team first confirmed that something in sewage effluent was making fish change sex. They were coming to the end of a three-year contract with the Department of the Environment, so they prepared a report of their findings. "Effects of trace organics on fish" was duly submitted to officials in Whitehall, where it was stamped classified and sent down to registry. Here, in the echoing vaults of Whitehall, document FR/D 0008 was not going to interfere with the smooth running of government in the run-up to water privatization. Effectively entombed with some of Her Majesty's most secret files, it remained buried from public view, occasionally dusted over by the odd cleaning lady, but otherwise darkness hiding any inconvenient questioning light for at least two years. "We were told we couldn't discuss this with the press, or even at scientific conferences," recalls Sumpter. " 'Classified' means no contact of any sort at all.[4]

"We were definitely frustrated," he says. "We thought we had a rather interesting observation and a very unique one. There was no report of this anywhere in the world before, and we thought this could be important. As scientists, you want to get results out, so that other researchers can be aware of the work and start studies in the same area. You make much faster progress if more scientists are involved. So it was quite intense frustration."

They made innumerable phone calls to officials and soon discovered that, if a report is classified, there was only one way of getting it declassified. The minister himself has to sign the approval. They also uncovered one or two other interesting aspects to the research.

When the Department of the Environment was made aware of hermaphrodite fish in the late 1970s, they had commissioned research from another British university. "Those researchers had taken effluent from the River Lea, dried it down into a powder, and formulated it into rat food and fed it to rats," explains Sumpter. "These results were apparently submitted to a very high level government committee, chaired by Lord somebody or another. A baron, two dames and three lords, among others sat considering this data. But I've never been able to see a report. I can't get hold of a copy. I'm not even allowed to read it at the DoE." The deliberations of the great and the good on the significance of feeding human sewage effluent to rats were to remain a mystery.

Nonetheless, he did get hold of some information about earlier studies, but there were many deletions. "I find it strange that the organization that is funding me is unwilling to tell me all the aspects of the story. And I think this is very common sadly. . . . Although to give the Department of Environment their due, they did fund us for six years and we would not have obtained our results if we had not received their support."

The machinations of the British establishment, with its blue-blooded committees, and its approach to scientific data as though it were some sort of espionage from the Cold War, were not going to defeat the scientists. John Sumpter, who had a very different background—"My parents had neither a bank account nor a car, and I was the first in my family to go to university"—took an opposite view. "There's no benefit in doing science unless it is to become publicly available," he argues. "It is a shared resource essentially, which everyone is paying for." He politely persisted in his requests to get the information declassified.

Every time he phoned the department, it was the same courteous reply. "I'm sorry John, it's on the minister's desk. We're waiting for the minister to sign it."

Releasing data of such a sensitive nature did not appear to be a high priority. Several ministers came and went and he was informed the report was still waiting on the ministerial desk. "It's a very effective way of essentially doing nothing fast," explains Sumpter, "and that went on for a very long time. It took almost two years to get the project declassified, which was done toward the end of 1991, and when it was declassified the Department of Environment placed a report of the work in the public domain."

Quite how the machinery of Whitehall disposes of inconvenient results, who gives the orders and on what memos it is signed, remains unclear. Under pressure from the scientists to publish, there was a uniquely British way out of the dilemma. To produce a report but put it in a place where no one would read it. The final report of document FR/D 0008 was duly printed on Her Majesty's stationery and taken from the registry to the library of the Foundation for Water Research. This was officially open to the public, but no journalist would have any reason to go looking for it. It would be very difficult for anyone poring over the finer details of plasma vitellogenin levels in mirror carp and rainbow trout to recognize that this might be of interest to the tabloid press. Sir Humphrey would have been well pleased.

A Department of Environment spokesperson told *Horizon* in 1993 that the results were classified because they were worried about public concern. And well they might have been. The research raised some tricky questions. Cautiously, John Sumpter summarized the dilemma they faced: "This sewage effluent enters British rivers, and in some parts of the country, especially in the summer, the effluent from sewage treatment works is well over half the flow of the whole river. *The river's not quite neat effluent, but certainly 50 percent effluent.* So the substance is present in the rivers as well as the effluent. About 30 percent of drinking water in the country is taken from lowland rivers, especially in south and east England. This water, subsequent to treatment, goes off to the consumer and we were obviously interested in whether this substance or substances was present in the water that was going to the consumer.

"Once water is extracted from the river, it will go through a purification process," explains John Sumpter. "This process can be variable, depending on what substances the water companies are trying to remove. But it's definitely possible that substances which are present in river water could get through the treatment process and would be present in the tap water that went to the consumer. . . . Quite a wide range of man-made chemicals are detectable in drinking water and it's very difficult to remove these: some of the PCBs, DDT, some drugs that are widely taken by people, like aspirin and paracetamol, and also some metals and a variety of ions. A number of compounds that are used in large amounts by people are now detectable in the drinking water."

Given this concern, the next obvious test was to grow the fish in drinking water directly to see if they would become feminized. This, however, was not possible because trout in drinking water would die anyway, for another reason! "Rainbow trout are very sensitive animals. You can't hold them for long periods in drinking water because of the chlorine it contains," explains Peter Matthiessen. "So we had this difficult problem of trying to identify one or more suspect chemicals out of all the compounds found in sewage discharges."

It seemed an overwhelming problem. A joint team from Brunel and the Ministry of Agriculture laboratories at Burnham on Crouch was assigned the unpleasant task of analyzing sewage. There were many potential chemical candidates to consider, but they had no way of knowing whether their tests were sensitive enough to detect them. There were very few clues from official channels, so they searched through the literature to try to find out what substances had been identified in sewage that were estrogenic.

While on the trail, they came across another team who had a very similar result and who also had not published. At the Musée Nationale d'Histoire Naturelle in Paris, Professor Roland Billard had been observing strange changes in eels in the Seine. He was also working with the vitellogenin assay. To his surprise, he found downstream of sewage effluent, immature female eels were producing a vitellogenin response as though they were fully mature females. He, too, had arrived at the uncanny conclusion that there must be something in the River Seine running through Paris that was acting like an estrogen. Sometime later, John Sumpter was also contacted by a German student doing her M.Sc. She was studying the gonads of different fish in the River Elbe near Hamburg and was finding many hermaphrodite fish. Her results, too, were strongly suggestive of estrogen contamination in the River Elbe. It was beginning to look as though this might be an international problem.

Then, through a literature search, Sumpter's team realized that American scientists were on the trail of a similar problem but with very different species: alligators and turtles. This was the work of Professor Louis Guillette and his team from the University of Florida. "We were extremely interested in what they were doing," recalls Sumpter. "It was highly relevant to what we were doing." However, in this case, the Americans believed that they had tracked down the most likely cause.

The Case of the Disappearing Alligator

Florida, June 1991

The helicopter roared over the Florida swamps, its propellers making a strange flickering shadow on the lush vegetation below. Three airboats were skimming the lake behind the helicopter, their skeletal metal frames dark silhouettes against the glare of the gulf sun. Lake Apopka, one of the largest lakes in Florida, close to the tourist attractions of Orlando, spread below like some vast glittering jewel. A team of scientists from the University of Florida had been called in to investigate the declining number of alligators.

During the 1980s, local farmers in collaboration with scientists from the Florida Game and Fish Commission found that the population had crashed and the eggs were not hatching in the normal way. The American alligator, known as *Alligator mississippiensis,* had been listed as an endangered species in the early 1970s, but in most lakes, populations had gradually recovered once federal protection was put in place. So why should the population crash on this lake? Working with the Government Fish and Wildlife Service, Professor Louis Guillette and Dr. Timothy Gross began their research by looking for alligator nests, brown dried bundles of twigs and sticks which could be clearly seen from the air among the luxuriant tropical growth of rushes and swamp.

"A hundred meters to your left. That'll be nest number fifty . . ." Franklin Percival yelled into the intercom in the helicopter to the scientists on the ground. A hundred meters below he could see the airboat leave the water and skim over the rushes toward the nest. "It looks empty. Just put your flag in it." The radio instructions were drowned as the helicopter circled the site. Louis Guillette and Timothy Gross stopped the engine of the airboat some twenty meters from the nest. Protected only by jeans and boots they stepped out into alligator territory. Their feet sank into the dark, warm, marshy water.

Of all the creatures that might provide insight into the human condition, it would have to be the alligator. Wild alligators are shy but they can also be extremely dangerous. When fully grown, sometimes reaching up to thirteen feet long, they have a considerable muscle mass and it is only too easy to be thrown off balance handling them. They had inhabited the area since the dawn of history, these creatures with the look of the dinosaur about them, and here they were sharing the twentieth century with the Kennedy Space Center, which wasn't far away. The technical brilliance of modern science seemed another world from this unchanging, timeless, prehistoric swamp.

As the scientists approached the nest, the ground became firmer. It was a loose construction of twigs and vegetation, cunningly concealed in among the reeds. They realized they could relax. The female was not guarding the eggs and didn't seem to be near. It was soon apparent why. "Nest 50. Six eggs. Broken eggs in nest. Two crushed eggs," Louis recorded on his data sheet. The eggs were buried at the bottom of the nest, several inches long and a beautiful yellow-white coloration. They took the temperature of the eggs but it was clear they were cold. This nest had been abandoned for some time. By studying the banding of colors on the outside of the egg, they could assess the extent of embryonic development inside. It was clear that even the undamaged eggs were not viable.

These were difficult working conditions. Even drinking liters of Gatorade, the local drink specially designed for fluid replacement, your skin still felt scorched and your throat blisteringly dry from the sun and wind after several hours. The team would make the long trek back to Gainesville to the university laboratory with their data and their samples.

In the cool, dark laboratories, by holding a high-powered light to the eggs, a technique known as "candling," it was possible to see the extent of the damage. There was a dramatic difference between the eggs from their control lake, Lake Woodruff, a wildlife refuge in a relatively isolated area, and Lake Apopka, which was surrounded by agriculture and development. "A good egg will have a beautiful pink rosy appearance which means there's lots of blood flow. The embryo needs that blood flow in its membranes to stay alive," Louis Guillette explains. "In Lake Woodruff about 75 percent of the eggs would look pink with a good blood flow and only 25 percent or fewer of the eggs are bad. In contrast, on Lake Apopka, most of the eggs, around 75 percent are translucent when examined under the light. They have no blood flow. This means either the eggs were infertile, in which case no embryo developed at all, or the embryo died early on in development. A few years ago, as many as 95 percent of the eggs on this lake were dead."[5]

Even to the untrained eye, the translucent, damaged eggs were quite different from the healthy eggs, a pale yellow instead of the healthy pink of a live, growing embryo. In the light, the gray shadow of the undeveloped yolk could be seen inside. Louis Guillette was becoming suspicious that the hormone balance of the species had been disturbed. "In order to make a viable egg, both the uterus and the ovary have to be exposed to appropriate amounts of estrogen. If you have varying amounts of estrogen, either too low or too high, the animal can't make a good quality egg," he explains.

The eggs that were alive from both Apopka and the control lake were incubated in the laboratory. This was done with great care to control the humidity and temperature because sex determination in alligators is temperature-dependent: high temperatures produce males and low temperatures produce females. This is typically an all or nothing response, so that embryos are males or females and there are few intersexes. Several times a day the temperatures in the trays were checked to ensure a steady 30–33°C and constant moisture. Nestled in among the moss, the young juveniles that hatched were soon to provide another important clue. The alligators from Lake Apopka had a much higher mortality. For some unknown reason, only a few of these young reptiles survived the first few weeks. Blood samples revealed another important clue.

Tests were done to check if the genetic material was altered in any way and to give a good guide to the general health of the animals. They also tested for levels of male and female hormones. There were striking differences. The female juveniles from Lake Apopka were very different to young female alligators from other lakes. *They had elevated levels of estrogen, almost two times greater than normal females from the control lake. Male alligators also had elevated levels of estrogen, and lowered levels of testosterone, three times lower than normal males.*

There were also other developmental abnormalities. Under the microscope the scientists could see strange alterations in the structure of the testes. In normal testes from a healthy animal, the germ cells, the cells that produce sperm later in life, are lined up in the center of the developing testes, and can be seen very clearly. "But the animals from Lake Apopka would have reduced numbers of germ cells," explains Louis Guillette. "The cells within the testes were poorly organized and there were unique abnormal structures, bar-shaped cells we hadn't seen before and which were a complete mystery." It was clear that, with reduced germ-cell numbers and these abnormalities, the process of sperm production was going to be impaired.

There were also changes in the females that would be consistent with overexposure to an estrogen. "In a normal female you have one egg per follicle and one nucleus per egg," he says. "What we found is multiple nuclei in each one of the eggs and many eggs within a single follicle. The abnormal nucleus with too much genetic material will probably result in embryonic death. An increased number of eggs doesn't mean twins or triplets. It is more likely to result in poor embryonic viability."

By the age of six months other abnormalities became apparent. A normal male will develop a phallus and paired testes and produce hormones such as testosterone. But by six months of age it was clear that many males on Lake Apopka were intersex with varying degrees of feminization. "In its mildest form," says Timothy Gross, "the intersex male would have a phallus and testes, but when you took a blood sample, it was producing female hormones, estrogens rather than testosterone. So it would be slightly demasculinized. In more severe forms the animal might have a phallus, but much reduced in size, and the testes might show some abnormalities and changes to look more like an ovary. In the most extreme cases we would find animals with no phallus at all, which we would anticipate was a female. However, its gonads were definitely more like testes than an ovary. So there were several different types of feminization that could occur."

As the young alligators grew, Guillette and Gross became more astonished at what they were seeing. In nearly twenty years of studying biology they had not seen anything like this. "Firstly we are actually seeing very obvious external changes that can be seen by eye, like changes in phallus size and structure. Secondly, there are internal changes to the gonads which can be seen down the microscope. In many cases the testes look more like an ovary. Finally, by examining the blood, we find changes in the hormones. The males are full of female hormones and have reduced levels of androgens, or male sex hormones. This doesn't mean the males are turning into females. They are never going to be normal females. They are going to be abnormal males: infertile and unable to reproduce."

When they went out to examine the wild adult alligators on Lake Apopka, this is exactly what they found. "Almost 80 percent of the male alligators caught on this lake have some kind of abnormal phallus or abnormal penis. Mostly it appears the abnormality is small size. The average decrease was around 25 percent, but many were as much as a half or two-thirds reduced. We were very surprised that the things we were seeing were so dramatic. We were actually seeing *sex reversal. The animals were changing sex* during embryonic development," explains Guillette. "What we were seeing was so incredibly different it just hits you in the face. There was something fundamentally different about Lake Apopka, so we just had to come back out and continue studying this."[6]

Alligators were not the only affected species. Beneath the surface of the water the beautiful blue-green shells of turtles could be seen. This species too was affected. "As many as 25 percent of the turtles on this lake are intersex," reports Louis Guillette. "Just like the alligators, under the microscope their testes have components of an ovary as well as testes. What this means is that the animal cannot produce normal eggs or normal sperm. It will have reduced fertility. We're not finding many normal males on this lake."

The team became convinced that exposure to an estrogen was the cause. "Firstly because we have elevated levels of estrogens at birth," says Guillette. "Secondly because when the animals hatch they already have the developmental abnormalities which you would predict from overexposure to estrogens. The kinds of adverse effects that we are seeing occur in identical studies that have been reported on laboratory mice that have been exposed to estrogens or studies on humans exposed to DES."

Louis Guillette became very worried at what he was finding. "These studies have only begun to address what we believe to be a serious widespread threat to wildlife populations. You have to realize that the vertebrate body, both animals and ourselves, have basically two 'control' systems, the endocrine system and the nervous system, and the principles by which they function are very similar. We have somehow severely interrupted the functioning of the endocrine system. Everything that we are seeing in wildlife has an implication for humans. I believe we have the potential to have major human reproduction problems. Human populations and animal populations could well decline dramatically, especially long-lived species like ours. But it may take us a long time to find out how we've been affected."

But what about the cause? They found that the lake had been designated one of Florida's most polluted lakes. It had been contaminated ten years earlier with a major pesticide spill from a chemical company on its shores. The chemical contaminant released is believed to have contained as much as 15 percent DDT and its metabolites DDD, DDE and chloro-DDT. The area adjacent to Lake Apopka became one of the Environmental Protection Agency's Superfund sites during the 1980s, targeted for a major cleanup. The main breakdown product of the contaminating pesticides was DDE: a compound which can mimic the action of estrogen and have other hormonal effects.[7] Earlier studies reported DDE in alligator eggs in levels of over 5 ppm, levels which had been shown to be harmful to birds.[8]

"One of the interesting things about Lake Apopka is that it's environmentally clean. Although it was listed in the early 1980s as a severely contaminated lake, today it's believed that as far as chemicals are concerned, it's perfectly okay," says Guillette. "If you go out and take a water sample, the water sample says there isn't much pesticide there. But one of the scary parts about the estrogenic compounds is that they're stored in fat, and so they're probably being stored in all the animals that are out here, not in the water."[9]

For Professor Sumpter hearing this story for the first time in the labs of Brunel University, the parallels between the changes to the alligators of Lake Apopka and the fish in British rivers were striking. In both cases, quite independently, the teams' first suspicions were that the animals must be exposed to an estrogen. But in Florida, the chemicals thought most likely to be causing the problems were present in such trace amounts they couldn't even be detected in the water. It was only as they accumulated up the food chain that they could be identified in the tissues of the animals.

The British team too were suspicious of DDT and its metabolites. They had, of course, checked all the published data on levels of insecticides and herbicides, especially the breakdown products of DDT. But they found that the concentrations of these compounds in river water and effluent were not known and couldn't be easily determined. "The concentrations of DDT and DDE are scarcely measurable in British rivers. They are present in parts per trillion, which is ludicrously low and very hard to measure," explains Sumpter. "When you get down to that level of analysis, you start generating as many 'ifs and buts' as you resolve."

Back in the lab at Brunel, scientists analyzing the human sewage effluent were finding all manner of chemicals which are excreted by humans: atrazine, caffeine, toluamide, phenols, hexanol, hydroxytrimethylpentane. A whole series of complex organics were duly noted, classified and labeled. None of them seemed to be estrogenic. But there was another very obvious candidate: ethinyl estradiol, the main ingredient of the contraceptive pill. "We arrived at this as the best candidate really by a process of elimination," recalls John

Sumpter. "We simply couldn't find anything else that might be responsible."

There were a number of good reasons for suspecting the contraceptive pill. "Ethinyl estradiol is a synthetic estrogen hormone and we knew this was extremely potent," says Peter Matthiessen, "far, far more active than natural estrogen. Millions of women take the pill. It is excreted from their bodies and is likely to get into the sewage." What is more, the pill had been taken by women for over thirty years, a timescale which corresponded well with the observed changes in sperm counts. Some early research reported finding it in British rivers. It was possible to do some simple calculations, working out how many people take the contraceptive pill in a given area, how much they excrete and how much of this is likely to end up in the sewage treatment works. From these calculations the pill seemed a plausible culprit.

The tanks in Burnham on Crouch were duly cleaned out and prepared for the next studies. This time, the fish were to take the pill. Rainbow trout were exposed to a graded series of concentrations of ethinyl estradiol mixed in their water in the tanks. Once more, the amounts were minuscule, not too dissimilar to levels that they thought might be in the aquatic environment.

This time there was success. The fish produced an identical response to the fish in the sewage effluent. "We found that ethinyl estradiol is extremely potent," recalls Peter Matthiessen. "It will cause the vitellogenin elevation in male fish at concentrations down to as low as 0.1 nanograms per liter, so that's 0.1 parts in a million million, so it's an extremely potent compound. . . ."

There was just one fly in the ointment. "We have been totally unable to detect any ethinyl estradiol whatsoever in sewage discharges. It just doesn't seem to be there," explains Peter Matthiessen. "Although I wouldn't rule out that ethinyl estradiol is a contributory cause of the changes we've seen, our current thinking is that it is unlikely to be the causative agent."

They found that the pill is excreted by women in a modified form. It appears the substance is chemically altered in the body to a less potent form. Other chemical groups bind to it so that it can no longer fit the estrogen receptor and have an effect. So even if ethinyl estradiol were present in sewage effluent, it would most likely be in an inactive form. "However, more research will be needed before we can eliminate it from our inquiries," says Dr. Charles Tyler, "not least because there are signs that microbes and bacteria in the sewage works may be able to turn the inactive form back into an active form. They break off the chemical group that has been added, releasing energy that they can use." This frees the active molecule of ethinyl estradiol once more.

But, as yet, they had absolutely no proof that either the pill or pesticides could be causing the changes to the fish in effluent. After weeks and weeks of testing they seemed to have got nowhere. Were their tests too insensitive, they wondered. Were they even looking for the right thing? They appeared to be at a dead end.

True to character, John Sumpter continued to carry out weekly searches of the literature. In 1992, the computer connected to the Internet was at the end of a large chemistry laboratory. Bowed over the screen, he scarcely noticed as technicians finished their tidying and left for the day. Held totally by the intricate details of his science, he checked every possible source in case there was one they had missed. There must be other chemicals out there which are estrogenic and which are in rivers, he reasoned, and the chances are someone has identified them. Then, to his amazement, as he scanned, he suddenly saw an article which had only just been published. It described an accident in a laboratory in America. This had led to a totally new discovery. As the article downloaded into the lab, Sumpter could see at once that the accidental finding in Ana Soto's laboratory in Boston could provide a very important clue. This could be the breakthrough they needed.

CHAPTER SEVEN

The Increasing Cast of Culprits

Boston, 1991

Boston academic life has always enjoyed a high profile. The large dome of the Massachusetts Institute of Technology dominating the skyline by the Charles River and the much-filmed nostalgic charm of Harvard Square are both well known to the public. But in scientific circles, Tufts University, near Chinatown, is equally well known, a center for research for almost 150 years. In laboratories at the medical school, Professor Ana Soto and Professor Carlos Sonnenschein were tackling fundamental questions in breast cancer research.

"Breast cancer is a very, very complex disease," says Ana Soto. "We are studying a tiny but important part of it: how estrogens can make breast cancer cells divide and multiply." It has been known for years that a woman's lifetime exposure to her own natural estrogen is an important risk factor for breast cancer. The longer the exposure, the greater the risk. The timing of key events in a woman's reproductive life can increase exposure and this correlates with breast cancer risk. Women who start their periods at a younger age, who undergo menopause at a later age, and who've never had or breast-fed a child are at an increased risk of the disease. All these factors are thought to increase the lifetime exposure to estrogen. It may seem strange that the estrogen that women need for sexual development and reproduction may harm them by facilitating the development of breast cancer. However, it has also been shown that women who have to have their ovaries removed at a young age, and therefore have much lower exposure to their own natural estrogen, have a much lower risk of breast cancer.

Sonnenschein and Soto were studying the growth of breast cancer cells to try to understand the relationship between estrogen exposure and breast cancer. They were culturing a particular strain of human breast cancer cells in the laboratory, a strain known as MCF 7. "MCF 7 cells come originally from an American nun; I think her name was Sister Catherine. She had breast cancer which had spread into the pleural cavity, between the lungs and the wall of the thorax," explains Ana Soto. In 1970, samples of this pleural fluid were donated for cancer research. The breast cancer cells were isolated and grown in a laboratory. Although Sister Catherine died soon after this, the cancer cells that killed her have been cultured and studied in laboratories around the world for twenty-five years. There was one important feature to this particular strain of breast cancer cells: they were known to multiply only in the presence of estrogens.

About 50 percent of breast cancers are sensitive to estrogens and will grow faster in their presence. Soto and Sonnenschein wanted to know why. By the mid-eighties, after some years of study, they felt they were making some progress. "Our results suggested that estrogens may act by blocking the effect of a protein present in blood which inhibits proliferation . . . that in effect estrogens may interfere with the inhibiting action of an inhibitor." They were deeply engrossed in these new ideas which challenged conventional thinking on how estrogens act in breast cancer. But one day when they came into the laboratory to check on the cells, they found something which stopped these investigations outright. It was an observation that was to change the direction of their research.

To the untrained eye, on entering the laboratory there are large incubators where cells are stored and cultured in the bright pink culture medium: stacks and stacks of bright pink and pale yellow flasks are carefully labeled, measuring different factors that might influence cell proliferation. The human breast cancer cells themselves cannot be seen directly, only under the microscope. But the researchers could see at once there was something very wrong. "It was an accident," Ana Soto recalls. "Suddenly the breast cancer cell cultures that we were studying started to divide and proliferate as though they were in the presence of estrogens. But we were absolutely certain that we hadn't added any estrogens to the trays. The cells were not supposed to divide."' It was inexplicable.

They repeated the experiment and got exactly the same result: the estrogen-sensitive breast cancer cells were multiplying fast, although no estrogens had been added to the experiment. "We were very concerned," explains Ana Soto. "This stopped our work. It was so clear the cells were proliferating. *It looked as if the experiment was contaminated by something that could act like an estrogen.* We had to investigate."

Sonnenschein and Soto had worked together for over ten years and never faced this sort of problem before. Because of the exquisite sensitivity of the cells, they had rigorous procedures in place to ensure that no contamination could occur. This was more than a scientific curiosity; it was frustrating and baffling. They scrutinized every aspect of their work, searching for any possible mistakes. "Life went on. But we were very depressed by the development," says Ana Soto. "We didn't know what was going on. . . ." They repeated the experiment several times, each time dismayed to find the estrogen-sensitive cells multiplying fast.

After some weeks of testing they could find no flaw in their experimental procedure and began to suspect the equipment and materials that they were using. At the time it seemed totally unlikely that the laboratory equipment could be anything other than lifelessly sterile. But the tedious routine of systematically checking everything had to be endured.

Eventually, they made some progress. When they changed one of the plastic tubes used in the experiment, the effect disappeared. The mystery estrogen seemed to be leaching from the plastic tubes. They could hardly believe this result. Ana Soto and Carlos Sonnenschein were well aware that some *pesticides and industrial chemicals* act like weak estrogens. But no one had ever reported estrogens leaching out of *plastic*. "You just don't expect plastics to be estrogenic," said Soto. "If plastics are involved it is a totally different game.

"We contacted the manufacturers and they provided us with

several batches of their tubes," Soto explains. "We tracked down the problem to the raw materials with which the tubes were made. We could detect it down to a point in which they changed the formulation of the plastic." So what, they wondered, was in the new formulation?

By now officials at the university who liaise with industry were aware of their problems. A meeting was set up with the manufacturers. "There were three representatives from the company, who had come down from New York, Carlos and I, and the vice-provost for research at the university. We introduced ourselves, and Carlos and I made a presentation to explain what we had found. . . ." Over tea and coffee, in the elegant surroundings of the conference room at the Hilton Hotel in Boston, the scientists quizzed the officials about the new formulation. "The manufacturers agreed there had been a change in formulation, but when we tried to ascertain the composition of the rogue material, we were denied the information, because that is a 'trade secret,' " recalls Soto. "We realized that we couldn't persuade them to tell us. And at that moment we made up our minds that we would find out on our own, because we considered it very important to know what this estrogenic substance was."

Their laboratory was not geared up for such complex analytical chemistry, but they felt they had no alternative. The implications of their findings could be immense. After all, they reasoned, if this estrogenic substance was in one plastic, could it be in others? Were other laboratory tests being derailed by a false result? Could this even affect medical testing, giving people false results? They could have simply switched brands of tubes and returned to their original research. However, they felt the problem was far too important to ignore. The next steps took many months. First they had to isolate the unknown compound from the tubes, then they had to identify it. Then, chemists at the Massachusetts Institute of Technology confirmed the structure of the mysterious contaminant. It was a chemical called *nonylphenol*.

Nonylphenols are widely used in industry and in some domestic products. They are very versatile and have smooth flowing properties because they bind easily to lipids or fats, and so can be used in paints, industrial detergents, lubricating oils, toiletries, agrochemicals, and as an additive to plastics and many other products. They've been manufactured for over forty years and persist in the environment and accumulate easily in animal tissues. "No one ever thought that nonylphenol was estrogenic. It just wasn't suspected," explains Soto. "It was always supposed to be safe. Yet it's so widely used I'm worried about our exposure to this chemical. Estrogenic effects may inadvertently appear in humans."

To make matters worse, nonylphenol is just one in a large chemical family of related compounds known as *alkylphenols*. There are over one hundred of these, and several are widely used. They bought the commercial preparations that are available and began testing these. Soon it was clear that other members of this chemical family are also weakly estrogenic. Professor Carlos Sonnenschein and Professor Ana Soto published their findings in a report in *Environmental Health Perspectives* in 1991.[2]

"The whole thing has hijacked our life," explains Ana. "We can't go back and just do our normal science. The implications of this were so large that we had to find out what is going on. . . ." Soto and Sonnenschein are still tackling fundamentals in cancer research, but now they also have a large program investigating estrogenic chemicals.

In London, their report on the accidental discovery of the estrogenicity of some alkylphenols was to have a strong influence on the direction of John Sumpter's research in the autumn of 1991. He already knew that nonylphenols were present in effluent from treatment works in really quite high quantities and was immediately concerned. He downloaded the article and read it, oblivious to everything except the quiet revelation coming from the computer monitor.

"I thought, here is a new group of estrogenic compounds, completely unrecognized before to be estrogenic, which had been very, very widely used for a long time throughout the world. Here was someone claiming that a breakdown product of these substances was estrogenic! It was very exciting!" Sumpter recalls.[3]

He immediately did a much more extensive search through the literature to check on nonylphenol. In less than half an hour he found many publications reporting the presence of the substance in the effluent from sewage treatment works. "It was obvious that the substance was present in quite high concentrations in probably any domestic sewage and there was at least a possibility it was responsible for the effects we were finding in fish. I felt that it was quite likely that this was the chemical we were looking for."

He found that around 20,000 tons are used annually in Britain alone. A third of this is estimated to end up in our rivers and lakes. In fact, he soon discovered because the National Rivers Authority were concerned about the presence of nonylphenol in rivers, they had already carried out tests. They had found nonylphenol present in some rivers in concentrations of at least 50 micrograms per liter, levels which are considerably higher than many other man-made chemicals in rivers.[4] So, reasoned Sumpter, could nonylphenol account for the changes in male fish?

John Sumpter discussed his concerns with Peter Matthiessen. "I was very surprised to see Ana Soto's results," Matthiessen recalls, "because the actual shape of the molecule nonylphenol is nothing whatsoever like estrogen."[5] It occurred to them that although nonylphenol may be estrogenic to human breast cancer cells, it might not be estrogenic to fish. There was one easy way to find out. At Burnham on Crouch, the fish tanks were prepared once more and trace quantities of the substance were added to the tank water in a range of concentrations. After just a few weeks the answer was unequivocal. "At 50 micrograms per liter we found it was very estrogenic to the fish," Sumpter explains. "The male fish make very high levels of vitellogenin. The results mimic those which we observed when we placed fish in effluent."

In the autumn of 1991, after an eight-year search, it seemed they were in sight of the goal posts. Nonylphenol at that stage was looking like the most promising chemical candidate to be causing the changes in fish. But this raised another concern. If nonylphenol could account for the changes in fish, could it account for the changes reported in man, through drinking water?

"We tested for them in river water and they are most definitely present, so it's most likely that they're also present in the water taken by the water companies to go to the consumer," says John Sumpter. "And although water is chlorinated and filtered in the treatment works, it's unlikely that the treatment works are doing anything to the water to remove these compounds." He was very concerned. "Here was a group of estrogenic compounds, present in quite appreciable amounts in water, and likely to be reaching the consumer. Presently in Britain, we have no information at all on whether these compounds are present in drinking water. There's just no information."

In fact they soon found that the Paris Commission, the international body which sets standards for water quality, had already recommended that nonylphenol be phased out by the year 2000 because of its toxicity to aquatic organisms and its persistence in the environment. In 1993, *Horizon* asked the water companies whether they test for levels of nonylphenol in the drinking water. The Water Services Association replied: "There is no need and no requirement in the UK Water Quality Regulations to look for these types of substance. Nor are sufficiently sensitive techniques available. Hence routine monitoring has not been carried out." The water industry told *Horizon* that, although some of the modern filtration methods do screen out chemicals like nonylphenol, not all river water gets treated this way.[6]

For the team in Brunel University, it was extremely difficult to work out whether there was any evidence that these substances were in drinking water. This issue was clarified one day when Dr. Sue Jobling, in a routine on-line search of "alkylphenols" came across some new data. She went straight to John Sumpter's office and knocked on the door. "Have you got a minute? I've just come across this paper on alkylphenolic compounds in *drinking water*," she said. It was impossible to conceal the excitement in her voice. "It's the very information we're looking for."

"Let's have a look." He was preparing for a lecture, but he immediately pushed the papers to one side. They could see at once the report came from Rutgers University, in New Jersey. A team of analytical chemists had been conducting tests on drinking water to search out organic compounds with a new and more sensitive method: particle beam liquid chromatography/mass spectrometry. Their aim was to evaluate the sensitivity of their new technique which they hoped would identify substances which had escaped detection before, in parts per trillion. To do this they had tested 500 liters of drinking water. They found a staggering number of different chemicals in water, and identified over *twenty* different types of alkylphenols.[7]

Sue Jobling and John Sumpter scrutinized their data. The result did not surprise them; they were well aware that there are probably several hundred chemicals in drinking water at very, very low levels. Nonetheless it was worrying to see a result like this set out in black and white before them. They added up the concentrations of all the different types of alkylphenols. It amounted to 1 microgram per liter.

"At that level I started to feel concerned," Sumpter says. "I began to mull over in my mind whether they might perhaps be doing something to the consumers of the water, whether they might be affecting the fertility in men?" Soon after this, he made contact with Dr. Richard Sharpe in Edinburgh to explain their findings in fish and the chemicals they were concerned about. "It seemed at least possible that these compounds were having deleterious effects on us. This would, of course, have very serious repercussions, and so it seemed important to investigate this issue. We really need to be looking at these chemicals now," he thought, "not in twenty or thirty years' time when it could be too late."

Sharpe and Sumpter decided to collaborate. "We spent an awful long time on the phone talking, swapping ideas and trying to consider how best we could attack the problem," Sharpe recalls. The problem seemed immense. There was no easy way to design a study which would give any insights into whether these chemicals might be having estrogenic effects in man. There were many points to resolve. Firstly, even though the substance was estrogenic to fish, would it necessarily be estrogenic to mammals? Although fish live in a completely different environment to mammals, there seems to be no difference at all in the hormones that control reproduction: they have the same hormones, which have the same function. So it seemed highly likely that hormone-mimicking chemicals that affected the fish would also affect mammals. The question of dose was much harder to resolve. The fish tested in Ministry of Agriculture tanks had shown a significant response in weeks. There was at least a possibility that humans would be exposed all their lives, but to much lower doses. There was no easy way to assess this.

"Obviously we can't do experiments on man, though maybe the experiments are already going on out there, unwittingly," explains Sharpe. "But this introduces the next problem. How do we relate any data on animals to man? If we could show effects in animals we would have reasonable grounds for supposing that the effects will occur in man. Unfortunately the converse is not true. If you don't show effects in animals, that doesn't rule out that the effects might be occurring in man. . . . If you look at the increase in Sertoli cell number, that takes just two weeks to occur postnatally in a rat. In man that can take up to twelve years. It is obviously impossible to re-create that twelve-year exposure period."[8]

Eventually they worked out an approach. The most likely mechanism, they reasoned, whereby estrogenic chemicals could cause an irreversible reduction in testicular size and sperm production is by decreasing the number of Sertoli cells formed in the testes during development. The lower the number of Sertoli cells formed in the male in the womb and during childhood, the smaller the testes and the lower the sperm count of the adult. So they aimed to assess the effect of an alkylphenol in mammals early in life when their Sertoli cells are forming. To achieve this, in one study, newborn male rats had traced amounts of octylphenol in their drinking water for the first three weeks after birth. In another study, in addition to this, the pregnant mothers were treated for eight or nine weeks while carrying the pups so the pups were exposed before birth as well. In all cases, the estrogenic chemicals were added to the water in tiny quantities of 1 milligram per liter.

"We're trying to administer the doses at very low levels," explains Sharpe, "which might approach those levels to which humans may be exposed. Obviously if we found effects in rats under these conditions which are really very, very much shorter than would be the equivalent human exposure, that would suggest there might be a human problem." Richard Sharpe was convinced the results would be negative. It seemed unbelievable that these tiny quantities, given in some cases just for three weeks, could have any effect whatsoever. Already the press were aware of the research and there were regular calls to the laboratory.

But, remarkably, while this study was under way, scientists in California had accidentally stumbled across yet another estrogenic chemical leaching from *plastic*. This compound, too, could be important in human exposures. The result was so unexpected that it had changed the direction of the Californian research for some time. It was the kind of contamination problem that scientists dread.

Stanford University, California

In the heart of the Silicon Valley, in Palo Alto, near San Francisco, the beautiful campus of Stanford University can be found beyond a grand avenue of palm trees, the Mediterranean-style buildings and arches lined with exotic Californian blooms. It was here, at the medical school, that Professor David Feldman and his colleagues were working in a very different field of biology, over a decade earlier.

Feldman was intrigued by an evolutionary puzzle: when did a particular class of hormones known as "steroids," which include the sex hormones, first arise? At what point in evolution did this extraordinarily complex and intricate process of communication between cells develop? To try to tackle this question, they were studying a simple unicellular organism: yeast. "We went back in evolution as far as we thought reasonable," explains Feldman, "to a single-celled organism such as yeast, and started research to see if steroid hormone-receptor systems could be found." If yeast could be shown to have a hormone-receptor system, it would be an important finding. It was already known that yeasts could communicate by means of simple messenger molecules, rather like hormones. Feldman wanted to know more about these messenger molecules in yeasts.[9]

Working to the limits of chemical analysis in the early eighties, Feldman and his team found that proteins in the yeast cells they were culturing in the lab could indeed bind with estrogens. This suggested that the yeast cells might have some kind of primitive "estrogen receptor." If, they reasoned, yeasts contained an estrogen receptor, then it was at least possible that the yeast also produced a hormone which would lock into that receptor. For some time they were on the trail of the "yeast cell hormone." Eventually they thought they had identified it: 17β-estradiol, the same as in humans. "Yes, we did think we had found it," reports Feldman. "For a while, we thought there might be a yeast cell hormone!" In their paper to the prestigious *Proceedings of the National Academy of Sciences* in 1984 they described how the substance they had identified in their experiments acted just like 17β-estradiol: in a number of tests its profile was identical. From this they speculated in 1984 that hormones like estrogen must have originated early in evolution to be present in a simple unicellular organism like yeast and that these simple single-celled organisms had a hormone-receptor system analogous to that in the vertebrates.[10]

There was just one problem. Although the estrogenic compound they had identified was 17β-estradiol, in some of the experiments with different culture media there was *no estradiol, but they could still detect an estrogenic response. Something else was present.* To their concern, they found they could get this estrogenic response using just water sterilized under pressure in polycarbonate flasks. "We did the experiment the exact same way, but left the yeast out of the flask as a control. Sure enough! We found we got the estrogenic response even without the yeast!" In this case, the hormone-like chemical wasn't coming from the yeast after all. It appeared to be coming from the water used in the experiment! But the water was sterilized and heated under pressure. It should have been completely free of contamination. It was a most bizarre result.

Just as Soto and Sonnenschein had to test every piece of equipment used in their research, the team at Stanford faced the same problem to track down the mystery contaminant. Eventually they traced it to an unexpected source. The estrogen mimic was leaching from the plastic *polycarbonate* flasks used to sterilize the water for the experiment. They purified and isolated the substance. The contaminant with estrogenic properties was *bisphenol A.*[11]

When they looked up what was known about bisphenol A, the mystery became clear. Bisphenol A has a very similar molecular structure to DES, the very potent synthetic estrogen which was used to prevent miscarriage during pregnancy and which, in some cases, led to a rare vaginal cancer in the exposed daughters. "Both are biphenolic compounds," explains Feldman, "that is, they have two phenol rings joined together. The natural estrogen fits the receptor like a key in a lock. These are like 'pass keys,' similar enough in structure to fit into the estrogen receptor as well." Although DES is a very potent estrogen, bisphenol A is much weaker. "In our experiments, it's approximately 1,000- to 2,000-fold less potent than the natural estrogen 17β-estradiol," he says.

As they checked the literature, they soon found that bisphenol A is used extensively in the making of polycarbonate. Polycarbonate is a clear, strong and rigid plastic, which is believed to be nontoxic and one of the toughest of all plastics. These properties have made it a favorite in many applications, including packaging for a wide range of soft drinks, water, and plastic products used in hospitals and laboratories. "In simple terms, the bisphenol A is like an individual pearl on a strand and the whole necklace is the plastic polycarbonate," Feldman explains. "So the individual bisphenol A molecules are linked together into a long strand to make the polycarbonate plastic."

Just as the Boston team were discussing their results with the manufacturer of the plastic tubes, the Stanford team were doing exactly the same thing with the manufacturer that supplied their polycarbonate flasks. The manufacturer was aware of the possibility of bisphenol A leaching, and had developed a cleaning regime which dealt with the problem. But the manufacturer's tests could only detect bisphenol A down to levels of 10 ppb. The Stanford team had a more sensitive assay which showed that bisphenol A could exert estrogenic activity in levels as low as 2 to 5 ppb and that it was biologically active in organisms such as yeast even at this level.[12] "What you have to remember is that hormones are amazingly potent molecules," Feldman says. "They normally circulate at incredibly low levels, so low it is hard to even describe them in normal language. Yet, in biological systems, these are active concentrations and systems exist to pick up these signals."

In 1993 Professor Feldman and his team wrote a full report of their findings in the scientific journal *Endocrinology*. The practice of preparing culture medium or water "in polycarbonate flasks is a common laboratory procedure," he warns. "The effects of this source of estrogenic substances on experiments are unknown, but may be substantial and may be a cause of confounding experimental results." As for polycarbonate in drink containers and other products, Feldman is concerned. "I think the possibility of this being important for consumers may exist. I don't want to create panic here because the likelihood of substantial estrogenic effects is quite low. We found very small amounts of bisphenol A. In fact we had to do the experiment several times to confirm that it was there. Nonetheless, this is potentially important and should be studied further to be sure that polycarbonate products are not adding to the pool of estrogenic substances in the environment. At the moment we just don't know how significant this is."[13]

Feldman's paper published in the journal *Endocrinology* in 1993 was immediately picked up in Boston and London. In Boston, in Sonnenschein and Soto's lab, the team began discussions straightaway to try to assess routes of human exposure to bisphenol A. In London and Edinburgh, Professor Sumpter and Dr. Sharpe wanted to know whether bisphenol A also, like the nonylphenols, could affect the sperm count of mammals.

Discussions were all the more urgent, because, by extraordinary coincidence, during this same period, Sumpter's team had discovered a *third* major group of chemicals used in plastics could also act like estrogens: the *phthalates*. It seemed estrogenic compounds leaching from plastics were everywhere, but no one had noticed them before. There had been no reason even to go looking.

Spurred on by the surprise discovery that alkylphenols could act like estrogens, the Brunel team had taken no chances. It seemed to them entirely possible that other common chemicals found in sewage might also be estrogenic. After all, no one appeared to be testing this. "By chance we had a project student," explains Sue Jobling, "and, like all undergraduates, she had to complete a research project to get her degree. So she was asked to compile a list of the twenty most common chemicals in sewage and to assess their estrogenicity."[14] This information was not that easy to obtain. They went to the Department of Environment and tracked down published and unpublished reports of compounds identified in sewage. The final list included a number of common compounds: toluene, acetone, benzoic acid, caffeine and many others. In every search, there was one family of chemicals which appeared time and time again: the phthalates. They were incredibly common, possibly the most abundant man-made chemicals in the environment.[15] At Brunel, they began to buy commercial phthalate preparations to screen them for estrogenicity.

Dark brown bottles of different types of phthalates started to accumulate in a corner of the laboratory. The phthalates are a large family of some fifty closely related chemicals. In the tissue hood, tiny amounts of the compound under test would be added to their cultures of the fish estrogen receptor. The natural estrogen 17β-estradiol was also added. The idea was to see whether the chemical under test would bind to the estrogen receptor, and therefore reduce the binding of the natural estrogen to the receptor. This is a well-known technique called a competitive binding assay.

In a matter of weeks they had a clear result from the cell cultures. Some of the phthalates could act like very weak estrogens. The two most potent were butylbenzyl phthalate (BBP) and dibutyl phthalate (DBP).[16] "We simply couldn't believe it," recalls Sue Jobling. "Phthalates are really incredibly common chemicals. So I took the chemicals straight into Malcolm Parker's lab in central London to put them through other screens. I asked if I could test them." In the labs at the Imperial Cancer Research Fund they were geared up to do other estrogenicity screens, such as tests on the breast cancer cell line that Ana Soto was using. "It took just two months of commuting into London and we confirmed the result. In fact it's the quickest paper I've ever written."

"We were particularly concerned," Professor Sumpter explains, "because they are one of the largest groups of industrial chemicals and they're very, very widely used in almost all plastics. Chemicals that are used in such large volumes will be present in just about all environments. They are in the air. They are in the water. It's very difficult to avoid exposure to phthalates.''[17] He found that phthalates are produced in such large quantities because they impart flexibility to plastics. BBP is typically used in the production of vinyl floor tiles, adhesives, synthetic leather and other industrial products; DBP is more common as a plasticizer in food packaging materials, poly(vinyl chloride) (PVC) and many elastomers.[18]

There were now three families of chemicals used in plastics that were under suspicion. Firstly, the *alkylphenols,* which included substances such as nonylphenol and octylphenol, typically used as surfactants in industrial chemicals. Secondly, the *biphenolic* compounds, including bisphenol A, used in making polycarbonate. And now the *phthalates,* found in PVC and a wide range of plastics. The cast of possible chemical suspects was growing fast. This was rapidly ceasing to be a conventional whodunit which could be satisfactorily resolved with the rounding up of one chemical villain. Far from being a straightforward detective story, in real life the problem had become both more complex and more extraordinary. Only a couple of years ago, discussions had centered almost entirely on those infamous molecules, well known to the scientific community, DDT and some PCBs, and their close chemical relatives, found in pesticides and industrial products. Now it was becoming clear that, in addition to these, some ingredients of plastics, innocently disguised in many of the products in our daily life, may also be culpable.

It was also clear that the research under way testing the drinking water of mammals was already in danger of getting out of date. So far, studies had begun on some of the alkylphenols, such as octylphenol, to see what effect very low doses might have during development. But the new findings suggested they should also be studying bisphenol A and some of the phthalates. It seemed entirely plausible that these chemicals also may affect the sperm counts of mammals.

In Edinburgh, the research was expanded to include studies of bisphenol A and butylbenzyl phthalate. Just as with octylphenol, these were added to the drinking water of pregnant rats in very low concentrations, 1 milligram per liter. The aim was to try to match possible human exposures. This was difficult, since so little is known about our exposure. Certainly, says Sharpe, the doses were much lower than in most toxicology research and given for a very short period of time, sometimes as little as three weeks. As before, the male pups' exposure was via the mother, during both pregnancy and feeding. Once weaned, no chemicals were added to the water. When fully grown, the male rats were examined for any adverse reproductive effects.

By the summer of 1995, the results became clear. They found that all three of these chemicals used in plastics, octylphenol, bisphenol A, and butylbenzyl phthalate, had significant effects on the offspring. Testes were 5 to 13 percent smaller and sperm counts 10 to 21 percent lower. In a paper to *Environmental Health Perspectives,* they wrote that, although there was no direct evidence of a link between human exposure to environmental estrogens and falling sperm counts in men, the study provided preliminary indirect evidence that exposure of rats to certain estrogenic chemicals . . . can result in reduced testicular size and sperm production. Since these effects occurred in less than nine weeks' exposure, whereas in man there is a corresponding window of development of several years, "there is at least a theoretical possibility that similar effects in men might be of larger magnitude than those described here for the rat."[19] While the scientific paper was being peer reviewed and the Medical Research Council publicity office prepared for the response, the press just couldn't wait. In the three or four years since Skakkebaek first announced his findings, Sharpe found himself continually in demand by media. In October 1995, the *Sunday Times* announced the results ahead of scientific publication: "Chemical link found to low sperm counts." This was soon a worldwide headline.[20]

In an interview for *Horizon,* Sharpe explained: "It was rather surprising to us, because we were adding these chemicals at very tiny amounts, 1 milligram per liter, in the drinking water of the mothers of these animals. So we anticipated that really there would be no effect and that we would prove that these chemicals were without adverse effects. Consequently the findings have surprised us. And, of course, with these very small effects we've had to repeat the experiments several times to make sure they are real effects, and that is what we've found: they are *real*."

Every time the tests were repeated, they obtained a similar result: the effect was highly reproducible, with an average decrease of 7 percent overall in testis size and 12 to 13 percent in sperm counts. John Sumpter was equally concerned. "This seems to me a very significant result," he told *Horizon*, "because it is the first time, I believe, that it has been shown that low-level exposure to estrogenic chemicals in the womb can lead to reduced sperm production when that animal reaches adulthood. This seems to me to be getting reasonably close to being able to demonstrate that if humans were exposed to similar levels of these chemicals, this could reduce sperm output. It is not possible, and rightfully so, to do the experiments on humans, so this is about as close as we can get to demonstrating the hypothesis that exposure to estrogenic chemicals affects the testes and sperm production. This seems to me at the moment to be about as close as you can get."

"What this raises is the *possibility* that if man is exposed to similar amounts of those chemicals to which we've exposed our animals, then we would have to anticipate that similar effects might occur," Sharpe confirmed.

The significance of what they were saying seemed all the more powerful because of the seriousness of their tone, Sharpe with his natural authority, which stemmed at least in part from his unwillingness to be drawn into overstating any scientific data, and Sumpter with his rigorous attention to detail and accuracy. Yet both were concerned that there was a reproducible and consistent effect which they had shown in data collected over two years.

The obvious next step was to try to assess human exposures. Could they in any way match the doses to which the rats were exposed? In Britain, Europe and America, tests were already under way to try to work out human exposures. Unknown to either Sharpe or Sumpter, research by Britain's Ministry of Agriculture and other scientific teams had already identified some of these chemicals in our food, in our food wrappings, and even in baby milk and bottles. The scale of the problem was about to unfold.

CHAPTER EIGHT

Routes of Exposure

For the scientists at the forefront of the research, assessing levels of human exposure presented insurmountable difficulties. "There are over 100,000 man-made chemicals in everyday use, and more are added each year," explains Professor John Sumpter. "We don't even know yet how many of these are estrogenic. Then there's the difficulty of knowing the routes of exposure. Initially we were very concerned about drinking water, because of the effects we had observed on hermaphrodite fish in rivers. But more recently, since the discovery that a much wider range of chemicals are estrogenic, it's become clear the potential routes of exposure are much greater. *Some of the compounds are volatile, so we're exposed through the air that we breathe. Others are in the water that we drink, such as pesticides. There are many in the food that we are eating and they are also in products that we touch and bathe in. So we have all these various routes of exposure and it is very difficult to work out which ones are the main routes.*"[1]

Although we can't see it, the evidence suggests our senses greet a very different world from that of our forebears. We are touching, eating, drinking, breathing and bathing in trace quantities of chemicals that can weakly mimic our sex hormones, many of which did not even exist a hundred years ago. In effect, some claim, our molecular environment is quite transformed. While initially these counterfeit hormones were thought to be in relatively few agricultural and industrial compounds, the realization that they could be found in plastics triggered a new wave of experiments. As scientists attempted to track down all the different routes of exposure, they found estrogenic chemicals, sometimes in quite large quantities, in products that were in daily use, a finding nobody had anticipated.

Plastics: Routes of Exposure

Granada, Spain, 1994

By chance, in the late 1980s, when Carlos Sonnenschein and Ana Soto were struggling to identify the estrogenic chemical contaminating their research, they were joined in the laboratory by a cancer specialist from Spain, Professor Nicolas Olea. Professor Olea had won a prestigious Fulbright scholarship, which enabled him to pursue his special interest in hormones and cancer with the Boston team. He became very concerned at their discovery that nonylphenol and related compounds were estrogenic. When, soon after this, Professor Feldman in California announced that another estrogenic contaminant, bisphenol A, was leaching from plastic in his experiments, Olea became even more concerned.

He discussed these results with his sister Dr. Fatima Olea, who specialized in food toxicology. She wanted to know whether there were other products which might contain bisphenol A and may enter the food supply. It wasn't long before they realized that, in addition to polycarbonates, bisphenol A is present in other plastics, such as those used as coatings for metal packaging, to line the inside of food cans, as bottle tops and for lining water pipes. Many metal tins used in canning foods and milk have plastic linings which are scarcely visible, unless colored. The lining is added to prevent metals contaminating the food or altering the taste. But, they thought, was it possible that the protective plastic lining could leach bisphenol A into the tinned food?

When they returned to Spain they decided to test this further. The University of Granada is nestled in the foothills of the snow-capped Sierra Nevada close to the Moorish old town and the Alhambra. In this magical setting, the Spanish team began a series of practical experiments which produced the most startling results about human exposure. They tested many brands of canned vegetables and milk. The cans were bought in Spain and the United States, but originated from a number of different countries, including Brazil, France and Turkey, as well as the United States and Spain. The liquid from the tin cans was filtered and tested for the presence of bisphenol A. The estrogenic action of this liquid was also tested by seeing whether it would make estrogen-sensitive breast cancer cells in culture proliferate.

"We got a very worrying result," reports Professor Nicolas Olea. "We found that in fourteen out of twenty cans, the contents of the cans contained bisphenol A in sufficient quantity that they could make breast cancer cells divide. Some 70 percent of the cans we tested showed activity in the test for estrogenicity."[2] The liquid from tinned peas produced the most marked effect. On average, the cans of peas each contained 23 micrograms of bisphenol A. The fluid from tinned artichokes, beans, mixed vegetables and corn also contained bisphenol A, in quantities of 10 to 18 micrograms per can. These, too, could make the human breast cancer cells divide and multiply in the laboratory.

When they filled the empty cans that had leached bisphenol A with distilled water and heated them under pressure, this water also contained bisphenol A. This confirmed their view that the contamination was due to the chemical leaching from the plastic lining of the cans. Tests on fresh vegetables showed no such effects. In their report of this study they pointed out that cans "are usually stored for months before they are sold. This prolonged period may favor the accumulation of bisphenol A. . . . *Findings such as ours demonstrate that humans are exposed and at risk.*" They urged for closer scrutiny of packaging materials to avoid such inadvertent exposures.[3]

In a statement in response to the publication of this data, the Association of Plastic Manufacturers in Europe confirmed that their studies by a multi-industry task group also show migration of bisphenol A from food and drink cans with these "epoxy-based linings." However, they point out that "migration levels are below the detection limit of five parts per billion for drink cans and an average of 63 parts per billion for food cans." These results, they calculate, indicate a low daily dietary intake, well within safety limits, of "9.6 micrograms per person per day. . . . In the light of existing knowledge," they conclude, "epoxy-based coatings are considered to be fully acceptable for the lining of food cans."[4]

The scientists are less sanguine. "We concluded that this represents a threat to the well-being of consumers," says Professor Carlos Sonnenschein. "The contents of the cans did increase the proliferation of breast cancer cells in culture. If this is something that happens in the laboratory setting, one can easily extrapolate that it may happen in the consumer. I think it is fair to conclude that these cans contain an estrogen mimic that may get into the blood, saliva and body fat of the consumers."[5]

Meanwhile, the Oleas had contacted other departments in the University of Granada to find out if bisphenol A was used in other products. At the School of Dentistry, they were informed it was used routinely. Most sealants designed to protect children's teeth from decay and many of the resins used in new white fillings contain bisphenol A. Dr. Rosa Pulgar, at the University of Granada School of Dentistry, became sufficiently concerned that she devised an experiment to test human exposures. Eighteen patients who were being treated with a standard amount of 50 milligrams of sealants had their saliva collected one hour before and one hour after treatment. The saliva was then tested for the presence of bisphenol A.

"We found a significant amount of bisphenol A was released after the procedure," reports Carlos Sonnenschein. Every patient treated with the sealants containing bisphenol A had the substance in their saliva one hour after treatment. The levels found in their saliva were high, between 90 and 930 micrograms, much higher than the levels found in the vegetable cans.[6] "This was worrying indeed because it meant that significant amounts of estrogen mimics got into the saliva and eventually would be taken up in the blood. . . . We found both the composite and the sealant has estrogenicity and we were certainly worried because this was yet another way humans might be exposed."

They found that even six months after treatment bisphenol A was still leaching into the saliva. "The oral environment is different in each one of us," says Soto. "You have to remember that fillings and sealants don't last forever. After about five years they are gone, worn down mechanically by chewing and grinding, and by the acids in the mouth. That means those 50 milligrams of sealants used, will sooner or later, get into your system."[7] The team also found that another component of composite resins, bisphenol A dimethacrylate, was estrogenic.

They reported these results at a conference in Copenhagen in June 1995. Dr. Sue Jobling from Brunel University was there. "I remember the slides going up," she recalls. "The levels really were quite high. I thought, I wonder if he's going to publish that!"[8] They did publish, and other scientists too were concerned at the levels. "The study showed up to 900 micrograms of bisphenol A could leach. *Micrograms!*" said Professor vom Saal. "It's a huge amount. In our studies with bisphenol A we're getting effects at 2 billionths of a gram per gram of body weight."[9]

Since then, the Spanish team have been contacted by over fifty dental associations who were very concerned. The American Dental Association in a position statement wrote: "This study, while well designed and executed, cannot be used to make any conclusions regarding adverse health effects attributed to leached components from dental sealants. Further tests and more clinically relevant experiments would need to be performed before any definitive conclusions can be drawn from these results. . . ."[10] Dental associations in a number of countries have started to investigate further.

The Spanish team also tested plastic baby bottles. "We found out that many baby bottles are made of polycarbonate and so we tested eight brands," explains Professor Nicolas Olea. In a preliminary and as yet unpublished study, they filled the new bottles with distilled water which was not estrogenic in laboratory tests. The bottles were sterilized for 20 to 30 minutes at 130°C and then the water was analyzed for the presence of bisphenol A. "Seven out of the eight brands produced bisphenol A," says Olea, "at levels of 10 to 20 micrograms per bottle. It was not a surprise to get this result, because we knew the bottles were made of polycarbonate." They presented these results at an international conference on food safety at Kaiser-Lautern University, Germany, in October 1996 and are continuing their tests on baby bottles. "It is likely that the baby does get bisphenol A from the bottle, but the amounts will be low," says Dr. Olea. "This has to be considered in the context of the baby's total exposure to estrogenic chemicals."[11] Earlier studies in Olea's lab have shown that some baby formulas may contain bisphenol A if they are packed in tin cans with a plastic lining that can leach bisphenol A into the powdered milk.[12]

In short, these initial studies show that bisphenol A is in several food products we consume. Although bisphenol A is only weakly estrogenic, nonetheless in the laboratory it has been shown to have sufficient potency that it can affect biological systems, such as in the multiplication of breast cancer cells. "The problem is how to measure the amount of total exposure to environmental estrogens," says Soto. "We may not be getting such a tiny dose of bisphenol A as we are of PCBs and other estrogenic chemicals. . . . We need to develop an assay so that we can measure all these environmental estrogens in blood. We would very much like to know what is in the blood of a normal person." At present there is little published data about the fate of bisphenol A in our bodies: how it is transported and metabolized and how quickly it is excreted. Consequently it is difficult to calculate the risks from human exposures. "Certainly we are worried," reports Professor Nicolas Olea. "We have only been working on this topic for three years, and we have already found three or four different ways in which we may be exposed to bisphenol A."

Professor Feldman, whose original studies at Stanford first raised the alarm is also concerned: "Multiple opportunities exist in the consumer world for plastic containers to contaminate their food and beverage contents and thereby perhaps to cause exposure of the population to bisphenol A. For example, large water jugs containing purified water are made of polycarbonate. Reusable bottles for soda, beer and other beverages may be manufactured from polycarbonate. The packaging of various items for babies, including food and juice containers, baby bottles, and baby food warmers, all of which might be heated in their routine use, is commonly made of polycarbonate. . . . There are many other examples of plastics coming into intimate contact with items that people eat or drink. . . . The potential for bisphenol A to be leached from plastic and to cause estrogenic actions in the population is real, and we believe further studies are warranted to evaluate this possibility."[13]

London, England, 1995

But bisphenol A is not the only synthetic estrogen which we are swallowing.

It turns out the Ministry of Agriculture in Britain has been testing for the presence of *phthalates* in our food for over ten years. Although it was not known until recently that some phthalates could act as very weak estrogens, other research indicated certain phthalates could be testicular toxicants.[14] Our daily pint on the doorstep, baby milk, food from supermarkets: none of these products escaped the attention of ministry scientists in the Norwich laboratories who were aware that this family of chemicals is a common contaminant of our food.

Part of this contamination is due to background levels in the environment. Phthalates such as di(2-ethylhexyl) phthalate (DEHP) are some of the most commonly used "plasticizers" worldwide, since they help to make plastics flexible. However, when used as plasticizers, the phthalate molecules are not firmly bound to the plastic and can therefore leach out. Consequently, since they are volatile, they are readily released into the environment and have become a ubiquitous contaminant which enters the food chain.[15] "It is utterly unsurprising that these chemicals are present in just about everything," says John Sumpter. "This chemical has been made so it doesn't rapidly degrade, otherwise the plastic would collapse. But since the plastic must also be flexible and able to bend, the phthalate molecules mustn't be locked in too rigidly; they need to slip over each other easily. That means they can leach out much more readily than they might otherwise do. . . ."

But the ministry scientists found the levels in food were too high to have originated just from general background contamination. They realized that phthalates were also getting into our food another way: *through the wrappings*. Aluminum laminates, the washes that are used on aluminum foil, plastics, and printing inks on food cartons and packets can all contain phthalates, and these can leach into foods. Because, like so many of the estrogenic compounds, phthalates are "lipophilic," or fat-loving, the risk of migration is greater with high-fat foods such as cheeses, crisps, pies, chocolates, butter and yogurt—in fact most dairy foods. Once in our bodies, they accumulate in fatty tissue and have been found in the body fat of wildlife and humans.[16]

In 1994 the ministry scientists in Norwich tested a typical basket of groceries and found phthalates in sometimes quite high levels in a wide variety of foods. In their article in *Food Additives and Contaminants* they wrote: "Samples of cheese, butter and other fatty products varied considerably in their levels of contamination, the highest being cheese samples containing . . . 114 milligrams of total phthalate per kilogram of cheese."[17] Other studies confirmed high levels in many products: gravy granules in printed paper or board, up to 62 milligrams per kilogram;[18] butter and margarine, 47 milligrams per kilogram;[19] chocolate, up to 46 milligrams per kilogram;[20] and 16 milligrams per kilogram in meat pies.[21] The levels found appeared too high to be attributable to environmental sources. The MAFF team suggested that "aluminium paper-foil laminate packaging may also be a major contributor to the contamination in the UK."[22]

Richard Sharpe and John Sumpter have expressed concern about this since the levels we may be eating are not that far removed from the levels to which rodents were exposed in their study.[23] Male rats were exposed through the drinking water of their mothers to low levels estimated to range from 0.1 to 0.4 milligrams per kilogram per day of butylbenzyl phthalate.[24] Even though butylbenzyl phthalate is a very weak estrogen, with a potency roughly one hundred-thousandth that of estradiol, this exposure still resulted in sperm counts reduced up to 20 percent and testes 10 percent smaller. "Ingesting 50 grams per day of such butters [foil-wrapped butters high in phthalates] by a 60-kilogram woman would lead to an intake of approximately 40 micrograms per kilogram per day," they wrote, "which approaches the nominal intake values in the present study. As the levels of total phthalates in other dairy produce can exceed 50 milligrams per kilogram and there are many other possible source of human exposure to these compounds, the present findings suggest that further studies of the estrogenicity of phthalates should be a priority."[25]

Ministry of Agriculture officials report that average dietary intakes of total phthalates "are considerably below the tolerable daily intakes set for some of these chemicals. The Department of Health has advised that there are unlikely to be any health risks to consumers."[26] The "tolerable daily intake" (TDI) is an estimate of the amount, expressed on a body-weight basis, of a contaminant which can be ingested every day over a whole lifetime without appreciable health risk.

However, Gwynne Lyons, an environmental consultant who has prepared a report on phthalates for the World-Wide Fund for Nature, points out that it would be relatively easy for any consumer eating a diet of fatty processed and packaged foods to exceed these suggested TDIs on a particular day. From the ministry's own figures for packaged foods, she points out that just 200 grams of a certain meat pie would exceed the recommended limits; so would four large sausages, plus a generous helping of chocolate cake, and a large portion of vegetable burger mix.[27] She also points out that the recommended limits were set by the European committee before it was discovered that certain phthalates are estrogenic. In the light of the new data, she believes the TDI for butylbenzyl phthalate may need to be revised downward.

"As early as 1987, the committees that considered the MAFF studies were saying they needed more information on the reproductive toxicity of phthalates," remarks Gwynne Lyons. "It was especially noted that they needed more data on testicular effects of butylbenzyl phthalate. Government-appointed experts have noted the lack of adequate toxicity data, time and time again, but have not seen the need to act effectively to limit exposures to phthalates. So the question must be asked, why didn't MAFF thoroughly investigate these effects and really find the lowest levels that could cause effects? . . . The Ministry of Agriculture has been doing lots of tests on these substances, but seeing that it is a government department I think it should release fully, with named products, details of all the foods that it has tested and the levels of each phthalate that are found in those things. I don't see why the Ministry of Agriculture bends over backwards to make reassuring statements, rather than portraying the uncertainties."[28]

A spokesperson for the Ministry of Agriculture told *Horizon* that "there was no immediate cause for concern and therefore they do not think it is necessary to release the brand names of food products tested."[29]

However, only three months later, the ministry became caught up in accusations of a culture of cover-up and secrecy.[30] Still battling with Europe over how to handle the collapse of Britain's beef industry following data suggesting that a new form of BSE (bovine spongiform encephalopathy) may perhaps be passed on to humans through eating contaminated beef, the government became embroiled in yet another very public storm. This time it was over data showing phthalates were turning up in baby milk. "Tainted baby milk!" was the headline in the *Daily Mail*,[31] "Sex change chemicals in baby milk," reported the *Independent on Sunday*;[32] "The five million ton plastic peril in food" was the *Guardian*'s version;[33] and *The Times* with greater restraint wrote: "Doubts that linger over baby milk."[34] The press outlined secret meetings, sex-change chemicals, babies at risk and a cover-up, all rolled into one. Government officials found themselves knee-deep once more.

"Alarming levels of chemicals that could impair human fertility have been found in leading brands of baby milk on sale all over Britain," reported the *Independent on Sunday*. "Ministry of Agriculture officials have secretly met manufacturers to get them to trace the source of the chemicals. . . . Neither the Infant and Dietetic Food Association nor the ministry will reveal the names of the brands tested or the levels of the chemicals found in them. The association says the information is 'commercially sensitive.' "[35]

The ministry scientists had tested fifty-nine samples from fifteen different brands of baby milk bought in five towns across Britain. Nine leading brands were tested by themselves. Six others were mixed for analysis. In a report of this study, they state: "Phthalates were detected in *all* samples of infant formula that were analyzed."[36] They continued: "taking as a worst case, the infant formula with the highest concentrations of the two phthalates which may have estrogenic activity [BBP and DBP], the combined estimated intake of these two phthalates was 0.023 milligrams per kilogram of body weight per day. This is between four- and seventeenfold lower than the dose range said to cause minimal effects in rodents." The dose levels of butylbenzyl phthalate that the male rats received through drinking water in the Edinburgh study were in the range 0.1–0.4 milligrams per kilogram of body weight per day.

The press, however, were not so easily reassured. They put a different interpretation on the data. In *The Times* it was stated: "Butylbenzyl phthalate produced damage [in rodents] when consumed at levels of 0.0–0.4 milligrams per kilogram of body weight per day. The ministry's analysis shows that a baby fed on formula milk would receive 0.1–0.13 milligrams per kilogram of *all pathalates*—about the same as the dose of BBP which caused shrunken testes and reduced sperm counts in rats exposed in the womb and in the first few weeks of life. Whether such a dose is more or less dangerous than BBP on its own is not known. The levels of phthalates found in the formula milk varied between 1.2 and 10.2 milligrams per kilogram."[37] *The Times*'s interpretation is arrived at by adding together the total phthalates found in baby milk, some of which may not be estrogenic.

The government faced angry accusations as it refused to reveal names of brands and further details of research. Assurances that there was no risk fell on deaf ears in the wake of the problems that had emerged over BSE. The Consumers Association called for an inquiry. Parents marched on Downing Street. The British Medical Association and Royal College of Nursing called for full results of the tests to be published. Telephone lines to doctors and advice groups were jammed. Angry mothers were interviewed on television, desperate to know what to give their babies. By Tuesday 28 May headlines were reaching crisis point: "Tories are losing control of milk." Since, for the Civil Service it was an extended holiday, reporters had to content themselves with a shot of a rather mystified porter at the Ministry of Agriculture closing the doors tight against the press.

It was left to Richard Sharpe to handle the crisis. He, too, was on a day's leave and was giving a talk on biology at his daughters' school. This was brought to an abrupt halt as the headmaster hurried up to him at the front of the hall, and explained that there had been a number of urgent calls for him from the Medical Research Council. His wife had called to explain that he would have to go into the lab. More than twenty journalists had called the Medical Research Council head office that morning.

Sharpe drove to the lab and called the head office. "What shall we do?" asked the press officer. "We've had every news program you can think of on the line. We don't know what to say. Is there anything in this or not?"

"I'll deal with it," said Sharpe. "I'll do what I can. I'll speak to BBC main news and ITN and that's it."

The two news programs duly turned up at the Edinburgh labs within half an hour of each other. "I think the scares have no foundation," Sharpe told the reporters for television news. "I hope that is some reassurance to mothers who are bottle-feeding their babies, the last thing that mothers should contemplate doing is changing to something less appropriate." Sharpe was extremely concerned that in the panic caused by the presentation of the data in the press mothers might in desperation switch to less suitable products. He was also angry that his own data, which he felt encompassed sufficient unknowns that it should not be used to fuel a health scare, was being misinterpreted in a way that might create genuine health problems. "I was very concerned to defuse the whole thing," he later explained. "You have to see it in the context that this had been oversold and sensationalized. Nothing good could come out of that situation. Only bad . . . If I had said there is a little bit of cause for concern the whole situation would have blown up even more."[38]

In fact, as he explained shortly afterward in an article in the *Daily Telegraph,* he feels strongly that "there should be more accuracy, that is, more *uncertainty,* in public utterances on health scare issues. . . . To the public, science is about facts. Facts=certainty= 'the truth.' To a scientist, 'facts' are more flexible; they are pieces of data that can often be interpreted in more than one way. Most facts are not 100 percent certain. This is because, with time, a scientist knows a 'fact' will prove to be only part of the truth and not the whole truth . . . certainty is a rare commodity."[39]

The very public row over how to interpret the risk to babies from phthalates in milk highlights the difficulties of weighing up the uncertainties even when the amounts of chemicals consumed are known. Firstly, explains Sharpe, the whole scare was fueled by one study. "Whenever you do an experiment, no matter how well you do it, you want someone else to confirm it, because there are always things that can go wrong. Normally one would wish for independent verification of the results. . . . Secondly, we need to know whether rats absorb and metabolize phthalates in the same way as humans. It looks as though phthalates have to undergo a chemical reaction in the gut to convert them into a form which is estrogenic. There is some data to suggest this occurs more readily in the gut of a rodent than in man," he explains.[40]

There are many other unknowns as well. So far, only approximately twenty out of fifty phthalates have been screened for estrogenicity, so it is at least possible that others may act as estrogens. Only one phthalate, butylbenzyl phthalate, has been tested in mammals at very low doses at a time when Sertoli cells are dividing. The effects of others have not been tested at these low doses in the drinking water of mammals. Thirdly, there is little data on the fate of these chemicals in our bodies: how they are absorbed from the gut; how they are transported in blood; whether they are degraded and into what compounds; and whether they are excreted or stored in our bodies like DDT and PCB. As will be shown, these factors are critical in weighing up the risks from a chemical. Nor do we know whether babies have less protection in their handling of these chemicals. At present, much of this data needed to assess the risk to infants is not known.

For scientists at the sharp end of the research, this can be frustrating. "We've come to realize that these things are more and more complex and so we have to be more circumspect," says Sharpe. "One of the problems is people wanting 'certainties' before you know you can deliver. And you know that at the end of the day what you are going to deliver is going to be inherently unsatisfactory because it is based around a number of presumptions."

The difficulties of carrying out testing on phthalates have been highlighted by one source close to industry, who explained that testing is a nightmare: ironically, phthalates are now so widespread, they are very difficult to measure. Even with all the right equipment, full-time staff and no expense spared, they are only succeeding in testing eight samples a week for phthalates. This is because, at each stage in the procedure, everything has to be specially treated to ensure that there is no contamination from other sources.

The only point on which there does appear to be agreement is that it is desirable to reduce levels of phthalates in baby milk. "I don't think the press were so very out of line," says Professor Sumpter. "And if phthalates are in baby milk by accident, what on earth are they doing in there? If there is physiological data to suggest exposure to these might just be a problem, well then, let's reduce exposure, or get them out completely. That seems to me the only sensible way forward." Sharpe agrees with this: "The manufacturers will tell you butylbenzyl phthalate is only used for very restricted purposes, such as the manufacture of linoleum. So there is still the question of how it actually gets into baby milk, because it is not supposed to be used in food contact applications at all."

Following the health scare, scientists working for the Ministry of Agriculture are continuing research on baby milk formulas to try to find out why phthalates are there and how they can be removed. Several European countries are taking measures to phase out the use of poly(vinyl chloride) in an attempt to reduce levels of phthalate exposure. The EU Science Committee for Food will also be reexamining tolerable daily intakes of phthalates in the light of the estrogenic concerns; however these studies are likely to take some time.

Despite the difficulties in interpretation, these recent studies on estrogenic chemicals in plastic do show that baby milk alone may be contaminated in several ways. Firstly, preliminary studies by Soto and her team indicate bisphenol A can leach from the plastic bottles into the milk and also from the plastic lining of milk powder cans into the milk powder. Secondly, Ministry of Agriculture studies have established that certain phthalates can be found in the formula milk itself. Since human breast milk is arguably even more contaminated with estrogenic chemicals than bottle milk (see below) it would appear to be impossible at present to avoid passing on at least part of our chemical legacy to the newborn. From the emerging data, it seems whether bottle-feeding or breast-feeding, all babies will be exposed to aspects of our chemical heritage, and no mother, no matter how dedicated, or where she is in the world, will be able to avoid it. Whether or not these observations have any significance for human health has yet to be determined.

Pesticides and Industrial Chemicals: Routes of Exposure

While the debate about the risks from estrogenic chemicals in plastics was going on in the full glare of the media, quietly in the laboratories the list of pesticides and industrial chemicals that could add to the estrogenic pool was still growing.

At Tufts University, after accidentally discovering that nonylphenol was estrogenic, Professors Ana Soto and Carlos Sonnenschein continued to test chemicals to see if they would act like an estrogen and make breast cancer cells proliferate. "I wouldn't be worried if only one estrogen was present out there," says Professor Soto. "But I am worried because so far we have only tested a hundred chemicals of the thousands that are used, and several of them turned out to be estrogenic. You cannot predict from their structure whether or not they are an estrogen, you have to test them."[41]

In 1995, Soto and Sonnenschein reported that two more pesticides were estrogenic, dieldrin and toxaphene. These pesticides have been restricted from use in America for at least fifteen years. However, like DDT, they also are fat-loving and bioaccumulate, and have been found in many species of wildlife. Their tests also showed that endosulfan, a pesticide which is still widely used, is estrogenic.[42] "There may well be some more surprises out there," says Professor Carlos Sonnenschein. "We are exposed to hundreds of thousands of compounds and there are many more that are being introduced each year, literally thousands each year. So if no check is imposed on this output, we have probably just got to the tip of the iceberg."[43] To date, using the MCF 7 cells, the breast cancer cells that proliferate in the presence of estrogens, they have identified more than thirty different chemicals which are estrogenic.

The following chemicals have showed estrogenic activity in Soto and Sonnenschein's tests. This is not intended as a definitive list, but as a preliminary guide to estrogenic chemicals. Just because a chemical appears on this list, they warn, this does not prove that it poses a threat to man, but is intended as an indication for further testing.

PESTICIDES AND INSECTICIDES[44]

DDT and its metabolites; endosulfan; methoxychlor; heptachlor; toxaphene; dieldrin; lindane.

INDUSTRIAL CHEMICALS

Some PCBs; alkylphenols, such as octylphenol and nonylphenol; bisphenol A, and some related biphenolic compounds; phthalates, such as butylbenzyl phthalate and dibutyl phthalate; butylhydroxyanisole; o-phenylphenol; p-phenylphenol.

PHARMACEUTICALS

DES; ethinyl estradiol.

Professor Soto points out how incredibly widespread and useful these chemicals are in modern society. There are insecticides and pesticides used in agriculture and disease prevention; plasticizers, such as phthalates and bisphenol A, used in packaging and other plastics; surfactants, such as some of the alkylphenols, used in detergents; antioxidants, such as butylhydroxyanisole, which may be used as a food additive to stop food going rancid; o-phenylphenol is a disinfectant; p-phenylphenol is an additive in the rubber industry; PCBs were used in the electrical industry. Then, of course, there are the pharmaceuticals, such as DES. It is hard to believe that they should all turn out to be capable of interacting with the estrogen receptor, at least in laboratory tests, and harder still to envisage something that could pervade so totally all aspects of modern living.

But assessing which of all these sources could be important in human exposures is extremely difficult. "Food is probably a very important source of exposure," Professor Soto points out, "because, if you think about it, at each step in the cultivating and processing of food there are chances of it being contaminated by estrogens. Firstly, when fruit and vegetables are grown, estrogenic pesticides can be added at that stage, either by direct spraying, or by soil contamination in the past. Secondly, during food processing and packaging more estrogens will be added through the plastic wrappings and packaging. If the food is canned, estrogens may well be leaching from the coatings. Finally preservatives, antioxidants, may be added and some of these are estrogenic as well. So the problem is that, at each stage, estrogens may be added and the chances are that we will be eating food contaminated by estrogens."

"This is very sad because we have the knowledge and the technology to avoid much of this contamination," says Carlos Sonnenschein. "For instance, not all pesticides are estrogenic, not all plastics leach estrogens. So we could legislate to avoid this problem. However, at the moment these substances are not tested for and labeled on our food, and so consumers have absolutely no way of knowing whether or not their food is contaminated!"[45]

We do know, however, that many of these chemicals find their way into our blood, our tissues and especially our body fat, since they can be measured there; and it is this accumulation or lifetime's exposure that may constitute the greatest risk. Levels of DDT and PCBs have been particularly well studied for nearly thirty years. Perhaps even more worrying than adult exposures are the levels to which infants may be exposed. The first report of the presence of DDT in human milk was presented by Laug and his team in 1951, who found healthy black women in Washington, D.C., had an average concentration of DDT of 130 ppb in their breast milk.[46] As new analytical techniques were developed it became easier to detect contaminants. Soon, other pesticides were identified in human milk, such as dieldrin and heptachlor. In 1966, Søren Jensen, a Swedish chemist, was the first to identify PCBs in human tissues. Since then, many persistent chemicals have been found, including pesticides such as aldrin, dieldrin, heptachlor, many different types of PCBs, and dioxins and dibenzofurans, amounting to a very wide range of chemical contaminants.[47]

Dr. Allan Jensen, head of the environmental toxicology department at the Institute of Technology, Denmark, has carried out a detailed review of chemical contaminants in human milk. Storage of these chemicals in fat "may be considered a kind of protective mechanism, keeping the chemical away from critical targets such as the nervous system," he writes. Because it is difficult to get samples of fat, unless a patient is undergoing surgery, breast milk fat is frequently studied. The levels in breast milk fat are much higher than in blood. "There is a dynamic equilibrium between levels of organohalogen residues (such as DDT, PCBs, and others) in blood and adipose tissue or fat," he explains. "The concentrations in blood will depend on mobilization of body fat stores and current intakes of the chemicals." He notes a single meal that is contaminated may significantly raise levels in the blood for a while and that generally current intakes of the chemicals are most decisive for blood levels.[48]

By contrast, the higher concentrations found in human milk "reflect to a great extent the levels in adipose tissue (body fat), a reservoir which is built up during many years and mobilized during lactation." Indeed, he argues, breast-feeding is the main vehicle for excreting these substances in lactating mothers. "In principle, most chemicals can be transferred by lactation. . . . Some infants may acquire 'adult' levels of persistent organohalogens in their body fat after a few months of nursing, in spite of their rapid gain in weight."[49] Because breast milk is the most important route for excreting these chemicals, he reports "in general there seems to be a tendency toward a gradual decrease in levels of organochlorine residues in milk during the first six to twelve months of breast-feeding." For instance, one German study showed a 30 percent decrease in DDT levels in breast milk fat during the first three months of feeding. An Austrian study reported the same trend. Furthermore, he found many studies confirm that levels of DDT, PCBs, and other organochlorines decline significantly with the number of births.[50]

He reports that DDT and its metabolites, DDE and DDD, are the most widespread contaminants in human milk. Average levels are now around 30 ppb of total DDT in whole milk, and higher levels of 1 ppm in milk fat.[51] In the West, following restrictions on the use of DDT, levels appear to be declining, but in some developing countries, where DDT is still extensively used in disease prevention and agriculture, levels can be up to a hundred times higher than this in some areas. The highest levels ever reported were in human milk from Guatemala in 1970 of more than 100 ppm of DDT in breast milk fat. This was found to be due to indoor spraying.[52] Interestingly, the levels of DDE and DDT still found in both breast milk and blood are up to a hundred times higher than the normal levels of the natural hormone estrogen circulating in women's blood.[53]

Recent studies on levels of PCBs in human milk also provide worrying results. At Reading University in England, Professor Raymond Dils and his team in the School of Animal and Microbial Sciences developed the technology to study individual PCBs. "PCBs are made by pumping chlorine through biphenyl to make polychlorinated biphenyls, or PCBs," he explains. "There are over 200 different permutations possible, depending on how chlorinated they are: that is how many chlorines attach to the phenyl rings. What we wanted to know was which of the PCBs were accumulating." In their initial studies on fish, they found that it was the more chlorinated and toxic PCBs which tended to accumulate. This prompted them to investigate human milk as well.[54]

They tracked down ten nursing mothers in the Reading area who were able to donate breast milk. Tests on this milk provided some worrying results. "Firstly, we found *sixty* individual PCBs in the milk. So of the eighty that are common in the environment, the mothers had accumulated some sixty of these," says Professor Raymond Dils. The concentrations were even more worrying. "The average concentration was 1,200 micrograms of total PCB per kilogram of milk, which is at least twentyfold in excess of the levels that the World Health Organization had shown to cause adverse effects in man." Finally, they found that twelve of the most highly chlorinated and toxic PCBs accounted for 80 percent of the total. These have been shown in laboratory studies to cause liver damage, carcinomas and adenomas. "I am concerned," he reports, "we just don't know the long-term effects of this; we just don't know what it means. We simply have no idea of the health risks for these infants in later life, and more studies should be done." He points out that the World Health Organization has recommended that mothers exposed to PCBs should not lose weight during lactation to minimize the risk of PCBs in fat being mobilized more rapidly and taken up by the milk.[55]

Interestingly, other scientists have arrived at a similar conclusion. Summarizing the risks from PCBs and dioxins and furans in milk, Dr. John Larsen writes: "Effects such as carcinogenicity, immunosuppression, and reproductive and behavioral effects . . . have been reported in laboratory animals at dose levels equal to, or only one or two orders of magnitude higher than, the levels normally found in human milk in many industrialized countries."[56] Allan Jensen, in his review of contaminants, concludes: "The safety margins are quite narrow. . . . PCBs and dioxins are considered to be a major concern and the most critical pollutants of mother's milk today, at least in industrialized countries."[57]

Overall, he finds that the persistent, bioaccumulative compounds in human milk "are normally 10–20 times higher than the levels in cow's milk or infant formula. This implies a much higher intake of such compounds by suckling infants compared to bottle-fed infants. . . . Nursing the infants may result in daily intakes of chemical contaminants far above the daily intakes of adults and above the internationally recommended acceptable daily intakes. . . ." Despite all this, Jensen still recommends breast-feeding. "Mother's milk is the natural and superior foodstuff for newborns and small infants; furthermore, nursing is of great immunological and psychological importance. . . . Virtually all national and international expert committees have hitherto concluded—on the basis of available information—that the benefits of breast-feeding outweigh the possible risks from the chemical contaminants present in human milk at normal levels."[58]

The Absence of Proof

However, even knowing that many of these chemicals are found in our saliva, blood, fat and milk still does not *prove* that they are necessarily doing us harm. Professor Stephen Safe, a toxicologist from Texas A&M University, America, has been one of the more outspoken critics of the estrogen hypothesis. For twenty-five years he has specialized in the study of such industrial chemicals as PCBs and dioxins, and argues that the whole issue has been blown out of all proportion. One of his objections is that the basic pharmacology does not add up. Most pesticides and other environmental estrogens are very much less potent than our natural estrogen, estradiol: they bind to the estrogen receptor, but much more weakly, sometimes *hundreds of thousands* of times more weakly. He has argued that tiny exposures to such chemicals are insignificant, dwarfed by our exposure to plant estrogens. The levels of estrogenic pesticides consumed per day, he calculates, could be as tiny as 2.5 micrograms. This should be compared with over a thousand milligrams of plant estrogens or phyto-estrogens.[59] He notes, for instance, that in Japan, where plant estrogens in foods like soya are eaten a great deal more than in the West, "men have a lower incidence of hormone-dependent cancers," and no evidence of increased incidence of reproductive disorders.[60] "We're eating a lot more natural estrogens in vegetables . . . than these trace levels of man-made estrogens. Until proven otherwise, let's not press panic buttons."[61]

Safe does acknowledge industry funding for some of his research. A footnote in Safe's 1995 paper to *Environmental Health Perspectives* states: "I gratefully acknowledge financial support for . . . research on dietary environmental estrogens from the Chemical Manufacturers Association."[62] When asked if this was the only paper in which he had industry support, he replied, "I've had some industry support over the years. They're funny, the way they fund. They'll support people whose research they don't agree with and those they do agree with because they like to be up on things. They don't like to be surprised. In my case what happened was, I gave a number of talks and presented a paper indicating I don't agree with this hypothesis and a friend of mine phoned me and said, hey, do you have some things you want to do because we'd like to support you. . . . Most of my work is supported by the National Institutes for Health and I've had EPA money . . . the chemical industry gave me support in 1994 and 1995 and there will be a number of papers coming out in which we acknowledge that support."[63]

This question of potencies and amounts of estrogenic chemicals consumed relative to plant estrogens has also been raised by the chemical industry. In an untransmitted interview with BBC *Horizon,* a spokesperson for the Chemical Industries Association in London told us: "The synthetic estrogens appear to be *very weak* estrogens—in fact they are about one thousand times less active than some of the natural products in our foodstuffs, the phytoestrogens, and a million times less active than the natural hormone estradiol."[64]

However, the principal objection raised by many scientists to such comparisons is that they fail to take account of what happens in real biological systems. While the laboratory tests can give a guide to estrogenic action and potency, they cannot reveal what happens to the compounds in the animal. There is mounting evidence that the fate of many of the man-made chemical estrogens in the body is very different from that of our natural estrogen, estradiol, or the plant estrogens.

Firstly, there is the question of how much is *absorbed* by the animal and gets through the gut. "There is evidence that some plant estrogens, such as genistein, don't get through the gut. They just aren't absorbed," says vom Saal, "as opposed to virtually all the man-made estrogens."[65] Most of the man-made estrogenic chemicals are soluble in lipids or fats. Because the membranes of the cell walls are made of lipids, this makes it easy for them to slip through and they are thought to be relatively easily absorbed.

Secondly, there is the problem of *transport* of the compound in the blood. The natural female hormone estradiol and, it would appear, some plant estrogens, do not normally circulate in a free form. They are bound in the blood to a carrier protein, called sex hormone binding globulin, or SHBG, until they reach the target cell. To interact with the target cell, they must first dissociate from the carrier protein, enter the cell, bind to a receptor and then interact with the DNA. The entire process is beautifully orchestrated so that concentrations of estradiol are strictly regulated in the body and it is not free to bathe all organs in its effects. "No plant estrogen or animal-made hormone lasts for more than a few minutes in the blood. There is no exception," says vom Saal.

By contrast, there is evidence that some of the man-made hormonal mimics bind much less readily, if at all, to carrier proteins like SHBG. "Many of the estrogenic chemicals haven't been looked at," says Richard Sharpe, "but those that have, *don't* bind to these proteins in the blood."[66] If the chemical circulates in a "free" form in the blood, in principle it is available to target and disrupt many cells. What is more, since they are lipid soluble they can pass readily into cells, just as they are readily absorbed through the gut.

"To give you an idea how this could affect potency," says vom Saal, "during pregnancy in humans, only about five in every 10,000 estradiol molecules are able to get into cells, and the remaining 9,995 are bound to proteins in the blood and they don't get into cells. So, if you find that in cell culture tests a chemical is 1,000 times less potent than estradiol, and yet it absolutely bypasses the barrier mechanisms in the blood, then you arrive at practically the same potency as estradiol." In recent studies with his colleague Professor Wade Welshons, he has checked out the potencies of some suspect estrogenic chemicals in breast cancer cell assays, but taken an additional step. Human blood was also added to see if the proteins in the blood would affect the binding to the estrogen receptor. "We checked how much the natural estrogen, estradiol is blocked from getting to the receptor versus how much of the man-made chemical can bind. We found we shifted the potency of these chemicals by hundreds of times relative to estradiol. What is more, these tests were much more predictive of what happened in animal studies."[67]

"Yes, there are differences in how they are transported in blood," concedes Stephen Safe. "But I would also caution you that there are all sorts of proteins which also bind to organochlorines (such as DDT or PCB) that we don't know enough about, so they may be bound up as well. We don't know. So their availability in blood could be a problem. . . . I agree with that." More research is needed on how these man-made chemical estrogens are transported in blood.

In addition to the question of binding, there is also the question of concentration in the blood. Some of the chemical estrogens have been found in concentrations a thousand times greater than natural estrogen. DDT, for instance, can be found in blood in levels of one part per million. Whereas a woman's natural estrogen will be circulating in blood at much lower levels, typically one part per trillion. Levels vary depending on the stage of the menstrual cycle.[68]

But the major and most worrying point of difference is the question of *storage* in the body. "Our natural estrogen, estradiol, is only free in the body for a matter of minutes . . ." explains Dr. Charles Tyler. "Its turnover is very, very fast. As soon as estradiol leaves its binding protein, other compounds try to attach to it, to render it inactive: sulphates, sugars and other things. So it has a very short time to get to the receptor, bind to it, and have its action. . . . It will be excreted soon after this. Ethinyl estradiol (the pill) may remain in the body for hours before it is excreted."[69] Plant estrogens are also metabolized quickly. There is no evidence they are stored in the body.

By contrast some of the synthetic chemical estrogens have the ability to remain in the body for months or years, building up a reservoir in the fat stores. "Metabolism and elimination from the body of persistent organochlorines are slow and in some instances apparently absent," writes Allan Jensen. "Most absorbed material remains unchanged, resulting in a gradually increasing *concentration* of the chemical substances in the organism (compared with the levels in the environment). Most are stored in adipose (fat) tissue, and their biological half-lives are long, of the order of several years."[70]

Stephen Safe does agree with this. "Yes, there is a big difference in storage. There are some environmental estrogens like the phenolics (bisphenol A) which probably don't have a long half-life, the same with nonylphenol: we don't know enough about them. But with the organochlorines (such as DDT and PCBs) we're talking about a long half-life—*infinite* compared with these other things. Phyto-estrogens, such as the flavonoids, are mostly flushed out quickly in a matter of hours; however, because we eat so much, there are measurable levels in blood. . . . But there is no question that there is a concern about the organochlorine estrogens in terms of their tissue persistence. I share that concern: that's why they are banned. . . . Nobody wants to touch them with a ten-foot pole and quite rightly so."

Nonetheless, they are in our bodies and passed on from one generation to the next. Sumpter and his team point out that, because of this accumulation in fat, organisms concentrate the chemicals to high levels, sometimes of the order of *micrograms* per gram of fat! This raises the real possibility that "chemicals that are weakly estrogenic *in vitro* (in laboratory cell culture tests) . . . may be active *in vivo* (in the animal) at considerably lower concentrations."[71] They point out that different chemicals are concentrated by organisms to different degrees. "For example, if you put a fish in a solution of nonylphenol at 1 microgram per liter," explains Sumpter, "after a few weeks you find that the concentration of nonylphenol in the fish is not 1 microgram per kilogram, but *300 micrograms* per kilogram. The fish concentrates the nonylphenol about 300-fold. Nonylphenol is said to have a bioconcentration factor of 300. DDT has a bioconcentration factor of 100,000!" What happens to these compounds once accumulated in an organism is essentially unknown.[72] They may be inactive when stored in fat, but when the fat is mobilized, as happens in reproduction—or even, perhaps, during dieting—the compounds may be freed to act elsewhere.

Summarizing these differences in absorption, transport and storage, Larry Hansen, from the University of Illinois, replied to Stephen Safe in a letter to *Science*: the very weak estrogens such as PCBs have 100- to 10,000-fold lower affinities for the estrogen receptors and are present in human blood at 2 to 8 nanograms per milliliter, whereas estrogen occurs in cycles in human female blood in amounts of 0.03 to 0.5 nanograms per milliliter and most of this is bound to proteins or disarming groups or both. Along with the PCBs are found DDE, mirex, methoxychlor, phthalates, and other pesticides and plasticizers in concentrations at least equivalent to those of endogenous estrogens. If the lesser binding and greater membrane transport of the weak estrogens are factored in, the pharmacology begins to add up.[73]

"If this is the only argument I had, I'd probably shut up," says Stephen Safe. "I understand the counterarguments and I don't dismiss them. . . ." Professor Safe does have other objections to the claim that certain chemicals are harming human health and these are discussed in Chapters 10 and 11. In addition to this, he adds: "My problem is this: it's a *hypothesis,* and you and others have presented it as fact, because it makes good news. . . ." However, on this point there is complete agreement. Sharpe, Skakkebaek, Sumpter, Soto and many others stress repeatedly that the case is far from proven and that it is a hypothesis, or theory, which is being explored. "It is not at all simple and straightforward," says Sharpe. "We can't say, we are exposed and therefore they have an effect. We can't even say, what is their potency relative to the natural hormone estradiol, and are we exposed to that amount? That's not enough because they may interact differently with different receptors in the body, we may have differing sensitivities to these chemicals. . . . Put all of these into the melting pot and add in a reasonable amount of unknown that we are going to find out . . . and you realize the difficulties of calculating exposures."

As is clear from the above, there are still a great many unknowns about the routes of exposure to synthetic chemical estrogens and the fate of these chemicals in our bodies. "But if we are going to study in detail for each chemical, its absorption, degradation and storage, we will never end up with an answer, not in fifty years," points out Professor Soto. "No one can tell you for sure about the risks until we run all the experiments. But with the knowledge that we have so far, do we really want to wait until we've run all the experiments?"

Furthermore, there is evidence that even if we had all this

knowledge about the fate of individual chemicals, this might still not be enough!

The Cocktail Effect

It was becoming clear that the problem was much broader than originally imagined. However, by 1994, yet another layer of complexity was emerging. It was looking increasingly likely that rather than considering the chemicals individually, their effects on each other needed to be taken into account. It was the *sum* of all exposures that needed to be calculated. We had to somehow *unscramble a chemical cocktail in our bodies*.

"One of the things that had always struck me at the Wingspread Conference was when I was listening to Tim Kubiak," recalls Ana Soto. "He explained how when you study wildlife, you always find the chemicals in combination. No one accumulates just one chemical, but a whole bunch of chemicals. That was always a problem. How were we going to assess that?"

They wanted to know how the chemicals would interact in biological systems. It was entirely possible that, in combination, chemicals could *cancel* each other out. Or they might act *additively*, so that the effects of two chemicals together equals the sum of the effects of the chemicals alone. Alternatively, they might even act *synergistically* with a multiple increase in potency in combination.

By 1994, Soto and Sonnenschein had analyzed more than sixty chemicals and found many to be estrogenic. They now had a collection of known estrogenic chemicals in the laboratory and could begin to test the "cocktail effect." By measuring the proliferation of human breast cancer cells in cell culture, they aimed to compare the strength of the estrogenic response of chemicals both separately and in combination. The greater the estrogenic response, the more the breast cancer cells would divide and multiply.

The trays of culture medium were prepared. Breast cancer cells were added at carefully measured concentrations: 20,000 cells per well. It was an experimental procedure they had done so many times before. However, the next step was to take minute quantities of selected estrogenic chemicals: endosulfan, toxaphene, dieldrin, tetrachlorobiphenyl, DDT, DDE, methoxychlor and others. Each one was added to a row of wells, at one-tenth of the effective dose. They knew these quantities were not enough on their own to produce any estrogenic response in the cells. The cells should not proliferate. However, in the final row of wells they pipetted ten of the chemicals together, each at one-tenth of a dose.[74]

Six days later they put on their lab coats and came to view the result. To any outsider viewing the scene through the laboratory door, it would be hard to appreciate the significance of what was going on. The laboratory looked as it always looked, bright, clean, with stacks of trays on one side, and brown bottles of the chemicals on the shelves. Huddled over the trays by the fume hood in the corner of the laboratory, the two scientists, both expert in the field of cancer cell proliferation, were totally engrossed in serious conversation. They were looking at trays which had now turned various shades of pink depending on the strength of the estrogenic response. Even before doing the exact measurements, it was clear from the color of the tray that the chemicals had acted together to make the human breast cancer cells grow.

"It was striking," Ana Soto recalls, "when we put the ten chemicals together even in these tiny doses, we obtained a *full estrogenic response*. We made the cells proliferate. In fact, in some cases we found that the effect is not simply additive; in a way they seem to have an effect that is bigger than the sum of its parts. They could be more potent acting together."[75]

They made their detailed notes, took all the measurements and prepared to write up the results of the experiment. "Intuitively we expected to find additivity," says Soto, "because all the knowledge we had indicated they could act through the same receptor. We did not think it was improbable. We ran it many times. We thought, 'That's fine. That's the way from our knowledge that it is supposed to be.' " But then they tried something else. Instead of testing ten chemicals at one-tenth of the dose, they tried five chemicals at one-tenth of the dose. In principle, this was not enough to obtain an estrogenic response.

"It was much more startling. Even with five chemicals at one-tenth of the dose, we were still getting effects," reports Ana Soto. "That was the moment when we began to think that the effect of the chemicals together could in some cases be more than the sum of the parts. Maybe in combination the effect could be synergistic, more potent when acting together. We do not have an explanation for this. We repeated it many times. We found that acting together, in some instances we could get an effect that was six- to tenfold greater than when acting singly. . . . It was very exhilarating, or at least very surprising. We started putting all this together and we thought what does this mean? It was so *ominous*."

At the same time in London, John Sumpter's team were also investigating interactions between chemicals. They were using a different technique, measuring the strength of the vitellogenin response produced by fish hepatocytes, cells from the liver which have an estrogen receptor. Once again, chemicals in tiny concentrations were added individually to liver cells: nonylphenol, octylphenol, o,p'-DDT, aroclor and bisphenol A. But when they were combined, the mixture of chemicals was considerably more potent than when each of the chemicals was tested individually.[76]

"When I saw the effects," reports Dr. Sue Jobling, ". . . I wasn't shocked, I suppose I was scientifically quite satisfied. It was the result I expected. My worry in setting up the experiment was that the chemicals would block each other's effect, and we would get a confused result. But this didn't happen. In fact each chemical added to the effect."

In a report of their findings, they warn: "In the 'real' world, organisms are exposed to mixtures of chemicals, not a single compound, and determining which chemicals are responsible for the alteration of hormone function from a complex mixture of chemicals is a difficult undertaking. In the 'real' world we have to consider not only the 100,000-plus chemicals in everyday use (and more are added annually), but also the products of their degradation or metabolism. . . ."[77]

Furthermore, their recent findings show that degradation products from one species may affect another. In a curious twist, their recent studies on vitellogenin production in male fish revealed that, in heavily populated areas of southeast England, the natural female hormones 17β-estradiol and estrone excreted by women are substantially contributing to the estrogenic cocktail in rivers. The female hormones are excreted in an inactive form and are somehow reactivated by enzymes in sewage works. While in densely populated areas the amounts present in sewage effluent could well be the key component of the cocktail to affect fish, in heavy industrial areas it is thought more likely that combinations of chemicals discharged into the river are the primary cause of the changes observed in fish in the wild.[78]

Identifying the cocktail of compounds in rivers affecting fish has proved difficult enough, and further studies are still under way. Unraveling the cocktail to which humans may be exposed, not just through water, but from food, air, and other sources, has become a puzzle of such unbelievable complexity that scientists are at a loss to know where to start. It had become clear that there was an ever-growing list of estrogenic chemicals. There was some evidence they had very different properties in biological systems in terms of transport, storage and activity from the natural hormone estradiol. Furthermore, it was looking increasingly likely that it was a chemical cocktail: acting together the response was more potent.

"Two years ago, we had what turns out to have been a very simple view of this whole process," said Richard Sharpe in a *Horizon* interview. "We now realize that it is much more complex. There are many more chemicals out there that are estrogenic—that is the first thing. Secondly, there are many more routes of exposure than we'd thought of. I think the most worrying development of all is the evidence beginning to come through which says that we can't consider all these chemicals individually but that their effects on us might actually be additive, so that we have to consider them all together. *Now if this is true, then we face a situation which is almost unparalleled because we really don't have procedures or the intellectual thinking that will allow us to evaluate the risk to man from this cocktail of chemicals and I think we face times of great uncertainty as to how we move forward. . . ."*

On top of all this, there was yet another extraordinary dimension to the puzzle. New research was revealing that it was not just a question of chemicals mimicking estrogen. *Some chemicals might interfere with other hormones in our bodies as well.* Estrogen mimicry was just part of a wider problem of "endocrine" or hormone disruption. The clues which led to this new perspective had come from studies of another chemical which had also turned out to be an endocrine disrupter—one of the deadliest poisons known to man, dioxin.

CHAPTER NINE

"A Universe of Chemicals"

North Carolina

"If we just focus on the chemicals that mimic estrogen, we run a serious risk of missing the broader picture," warns Linda Birnbaum, head of the experimental toxicology division of the EPA's Health Effects Laboratory. "It is important to realize that there is a universe of chemicals out there that may interfere with other hormone systems. There is evidence that the adrenal glands are one of the most affected and there is evidence of other hormonal effects as well. The entire endocrine, or hormone, system is so finely balanced, if you affect one part, it is likely to have an effect elsewhere in the system."[1]

For years, scientists had concentrated their efforts on developmental effects of estrogens because of the tragic cases of the DES-exposed women who developed rare vaginal cancer around puberty. However, pioneering studies by Dr. Earl Gray and others at the EPA labs in North Carolina were revealing that some chemicals might interfere with *male hormones and even other hormones in the body*.

The endocrine system comprises several main organs which secrete the chemical messengers, or hormones, into our bloodstream. These are transported to target organs in other parts of the body, with chemical instructions directing the cells how to behave. In effect they choreograph the delicate and finely tuned process which regulates the levels of substances within the body, a process termed "homeostasis." Everything is maintained in perfect balance. For instance, insulin, secreted by the pancreas, regulates levels of sugars in the blood and adapts them so they can be absorbed by muscles and other tissues that need to convert sugar into energy. Lack of insulin can lead to diabetes. Growth hormones regulate what happens to fats, carbohydrates and proteins in the body. The sex hormones, or steroids, control sex development and maturation.

There are several main organs involved in the endocrine system. The pituitary gland, at the base of the brain, is seen as the conductor of the entire orchestra. It has a great number of secretions which can influence the other key players, such as growth hormone. The thyroid gland, in the front of the neck, secretes thyroxine, which regulates the general metabolism of the body. Just above the kidneys are the adrenal glands which produce the stress hormones adrenaline or noradrenaline. Adrenaline is the "flight or fight" hormone, produced when the body is under stress, to dilate the arteries, increase blood sugar and give energy for a speedy getaway. The adrenal glands also produce cortisone and hydrocortisone, which are essential for life. The ovaries and testes produce the sex hormones estrogen, progesterone and testosterone.

"There is real cause for concern," says Dr. Earl Gray. "Our studies suggest that it is not just the estrogen receptor that is targeted, but other hormone systems as well. . . . I think we need to rethink the whole issue."[2] Dr. Gray's studies were to shed light on the mechanism of action of one of the most infamous molecules of the century: DDT.

Scrambling the Male Hormone

As a reproductive toxicologist, Dr. Earl Gray had detailed knowledge of how chemicals could alter reproduction. For many years his research centered on the mechanism of action of the familiar estrogenic pesticides and PCBs. But by chance, in the early 1990s, he came across studies which alerted him to the possibility that chemicals were disrupting male hormone messages as well.

During the pesticide registration process, the Environmental Protection Agency had received a curious set of data from a company that manufactures a fungicide called vinclozolin, which is sprayed on fruit and vegetables. Their animal studies showed that the fungicide had an unusual effect on the male fetus. Male rats exposed to vinclozolin appeared to demonstrate typical effects of estrogen exposure. But the females were completely unaffected. This result was odd and the data was sent on to Earl Gray to see if he could make any sense of what was going on.

"Their data was a big surprise," recalls Gray. "It clearly suggested we were dealing with some sort of 'anti-androgen' syndrome, which was 'demasculinizing' the males but leaving the females unaffected. Somehow the action of male hormones or androgens appeared to be blocked." Further studies confirmed that male rodents exposed to vinclozolin in the womb did indeed show delayed puberty and malformed reproductive organs, including undescended testes. Some males even developed a "vaginal pouch" and breasts more characteristic of the female, whereas the females suffered no estrogen-like alterations at all, such as effects on cyclicity or fertility.[3]

In laboratory studies, Earl Gray and his colleague Bill Kelce, together with Elizabeth Wilson at the University of North Carolina, were able to show that the fungicide had exerted its effect by interfering with the male hormone receptor, or androgen receptor. When they added testosterone to the male hormone receptor, it would, as expected, bind to it and activate the genes. But when they added testosterone *and* vinclozolin, the fungicide would compete with testosterone for the male hormone receptor sites and block them. Consequently, it reduced the activity of the male hormone, testosterone, by literally jamming the receptor sites. Just as estrogen mimics target the estrogen receptor and can fit it like a lock and key, this chemical was doing exactly the same with the male hormone receptor. The animal was feminized, says Gray, not by overexposure to female hormones, but by *underexposure* to male hormones. It was a different form of feminization. "We conclude," they warned in their article summarizing this study, "that similar alterations are likely to occur in humans if the fetus is exposed during sex differentiation to active metabolites of vinclozolin at levels equal to or near the tissue levels attained in our rodent studies."[4]

This work was soon to provide unexpected insights into the insecticide DDT. "The first studies showing DDT could have reproductive effects were in 1950, and for over forty years since then, no one really had a clue exactly how it exerted its effects," says Gray. It was presumed, since it had feminizing effects, that it was somehow acting like an estrogen, although the precise mechanism of action remained elusive.

In the spring of 1994, just as Gray's team had completed their study on vinclozolin, they were invited to a scientific symposium where Professor Louis Guillette of the University of Florida was also giving a talk. Guillette was presenting details of his remarkable findings on the alligators at Lake Apopka. "As the slides went up," recalls Gray, "showing the strange alterations to the testes, the reduced phallus and the altered hormone levels, including the near absence of testosterone, I suddenly thought, his alligators looked just like our rats exposed to vinclozolin!"

If this was correct, it would suggest that rather than the alligators being *feminized* by overexposure to estrogens, it was possible that they were being *demasculinized* by having their male hormones blocked in some way. The same thought had also occurred to Louis Guillette. It had already been shown that the main contaminant to which the alligators had been exposed was DDE, a breakdown product of DDT. Some of the alligator eggs had high levels, up to 5,800 ppb, of DDE. This would suggest that DDE was not interfering with the action of the female hormone, estrogen, at all, but with the male hormone.

Earl Gray and Bill Kelce began to discuss these ideas. "It was really a complete stab in the dark," Gray was to recall later. But if it was true, it would solve a long-standing scientific puzzle. Since Gray was due to give the next talk at the meeting on the effects of vinclozolin, they were forced to wait. "We gave the last talk, and then we literally ran out of there and back to the lab. We were in a hurry, really keen to test this. We thought it was a real, viable hypothesis that DDE was in fact an anti-androgen."

They tested DDE in a male receptor binding assay, just as they had for vinclozolin. Within a few days, they had preliminary results. "It was a pretty exciting moment," recalls Gray. "DDE *did* bind with the male hormone receptor. It was blocking the action of the male hormone."

This result showed that the action of DDT is much more complex than was previously thought. Commercial DDT is really a mixture of two chemicals, Gray explains. "About 15 percent of DDT is a chemical called '*o,p'*—DDT,' which is estrogenic in its action. The remaining 85 percent is '*p,p'*—DDT,' which has one chlorine in a different position. This confers very different properties, almost like a different chemical." *p,p'*—DDT degrades into *p,p'*—DDE, which they now knew could act like an anti-androgen. So, in effect, DDT has a double action on the sex hormones. One of its component ingredients can act like an estrogen; the other is an anti-androgen and blocks the action of the male hormone.

Further tests confirmed their preliminary findings. They were able to show that DDE prevents the male hormones from turning on the genes they normally activate. What is more, they found that DDE is a *potent* androgen blocker. "This is pretty extraordinary for a chemical that we used to spread in pounds per acre," says Gray. "We felt this was very important, because DDT has been used all around the world and bioaccumulates. It is persistent in many environments, and there is a lot of wildlife and human exposure."

A few weeks later they showed their results to Guillette. "He was giving a seminar at the labs across the street and we brought him over," Gray recalls. "Bill Kelce had plotted out graphs showing our results. He put these things out on the table for Louis to have a look at. Louis was just kind of 'holy cow!' I think all three of us looked like we had just swallowed canaries; we were really pleased. It was pretty exciting. It's not that often that you have a hypothesis which works just as you thought it would."

With Louis Guillette as "mentor," says Gray, they continued further studies and reported their findings in a letter to *Nature* in June 1995.[5] They were also invited to present the details of the studies at the American Zoological Society annual meeting. This work received a great deal of attention. "Now I can explain what

I am seeing," Guillette told reporters from *Science* later, "DDE should be viewed as an endocrine disrupter."

Gray and his team soon found that many synthetic chemicals with estrogenic activity can bind also to the androgen receptor, including some fungicides, DDT metabolites and other estrogenic chemicals. "The androgen receptor is extremely promiscuous in what it will bind to, even more so than the estrogen receptor," he says. "We're very concerned about these results, especially given the levels of DDE in human tissues. It has been reported to reach levels in human tissues that are equivalent to the levels that bind to the androgen receptor in our studies: between 60 ppb and 1 ppm. And there are plenty of studies of human milk where levels of DDT exceed these levels of 1 ppm. There may be all sorts of health effects from this."

What is more, it wasn't just interference with the male and female hormones that they were worried about. Evidence was mounting that the action of other hormones may be disrupted as well. The clues for this perspective came from studies of one of the deadliest poisons known to man: dioxin.

The Legacy of Dioxin

It is perhaps hardly surprising that dioxin should have a particular hold over public imagination. Laboratory studies have shown dioxin to be one of the most dangerous chemicals ever produced. It can be a thousand times more deadly than arsenic. Less than a millionth of a gram will kill a guinea pig.[6] Scientists discovered its unwelcome presence as a widespread contaminant through a series of unfortunate accidents.

The first clues about dioxin poisoning arose in the 1930s and 1940s among sailors. They developed a rare and very disfiguring skin disease called chloracne. A severe rash, like an "angry kind of acne" could cover the whole body and last for years and years. It was common practice for sailors to scrub the decks of the beautiful battleships with special waxes called "halowaxes." Unknown to anyone at the time, these were sometimes contaminated with trace amounts of dioxin-like compounds.

During the 1950s and 1960s a number of industrial accidents, often in herbicide plants, led to outbreaks of the same mysterious skin disease. Investigators began to examine the chemicals to which workers had been exposed. They identified one minor, but very toxic contaminant: 2,3,7,8,—tetrachlorodibenzodioxin, otherwise known as TCDD or dioxin. But still its pervasiveness as a contaminant was not recognized.

Then the tragedy really took off. At the height of the Vietnam war, the American government sprayed millions of gallons of herbicides over the lush tropical vegetation of Vietnam. The aim was to defoliate the rain forest and expose the enemy. "Agent Orange" was the key herbicide used, a potent mixture containing two herbicides, 2,4-D and 2,4,5-T. At the time, public concern centered on the use of napalm, which was used to burn villages and torch whole areas. The horrific injuries resulting from this were immediately apparent. Nobody realized that one of the herbicides in Agent Orange, 2,4,5-T, was contaminated with dioxin.

But when the veterans returned from the war, they began to report a number of serious medical problems, including cancer and deformities in their children. Then dioxins were measured in the soil of Vietnam where Agent Orange had been sprayed. Finally, it was discovered that 2,4,5-T could be contaminated with dioxin during its manufacture. The levels of contamination varied depending on the manufacturing process, but there was nearly always some dioxin present. People began to question whether dioxin exposure had contributed to the veterans' medical problems. Laboratory tests were beginning to reveal just how dangerous dioxin could be. At the Environmental Protection Agency, studies revealed that dioxin was the most potent carcinogen that had ever been tested.[7] After a decade of debate, 2,4,5-T was suspended as a herbicide by the EPA.

"In the mid-seventies, we were just waking up to how dangerous dioxin could be," recalls Linda Birnbaum. "There are more toxic *biological* agents, such as botulinum toxin, but as far as manmade chemicals go, this was the most toxic." Papers were produced confirming the harm it could do. In laboratory animals it could damage the thymus, spleen and testes, enlarge the liver, and cause cancer and birth defects.

Then, in July 1976, there was another accident. An explosion at a chemical factory in northern Italy created a cloud of dioxin over the city of Seveso. Thousands of people were exposed. Almost immediately there was a severe outbreak of the skin disease, chloracne. Further studies were started. These revealed very high levels of dioxin in some of the exposed residents, up to 56,000 parts per trillion.[8]

By now, advances in chemical detection made it possible to detect dioxin at extremely low levels: even less than one molecule in a trillion molecules of water. Suddenly, just as with PCBs and DDT, it became apparent that dioxins were everywhere. They were in industrial wastes and landfill sites. They could be detected around incinerators. Then they were found in our food, in the air, soil and water, and even in mother's milk. One health scare after another emerged as they turned up in a growing list of domestic products, even disposable diapers. Gradually it was realized that dioxins and their chemical cousins are the unwelcome by-products of many industrial and manufacturing processes. They have been found as a contaminant in the manufacture of certain herbicides and other chemicals such as PCBs and chlorinated benzenes and phenyls. They can be found in effluents and industrial wastes, such as those from the pulp and paper industry, the bleaching industry, and others. Most especially they are found as the by-product of combustion. They have been identified in fly ash and emissions from municipal, hospital and hazardous waste incinerators.[9] They have been detected in car exhausts and spent motor oils; levels are higher over cities. Then it was realized that there has always been a natural background level of dioxins, which can form whenever things burn. "However these levels were barely detectable prior to the 1920s," explains Linda Birnbaum. "In the last fifty years, as the chemical industry has developed, there has been a dramatic increase in the amount of dioxin present in the environment."

As chemical analysis improved, it was found that there are 75 different chlorinated dioxins and 135 chlorinated dibenzofurans, a close chemical cousin. Other related compounds include 209 chlorinated biphenyls, as well as many chlorinated diphenyl ethers, naphthalenes, azobenzenes and azoxybenzenes. Some estimates suggest there are potentially some 5,000 related compounds if all the chlorinated and brominated forms are counted.[10] However, not all of them are toxic. The most dangerous is 2,3,7,8-tetrachlorodibenzo-p-dioxin, known as 2,3,7,8-TCDD. Next in line are 1,2,3,7,8-PCDD and 2,3,4,7,8-PCDF, which are estimated to have half the potency of TCDD.[11] Several of the other closely related compounds have dioxin-like activity.

Like PCBs and DDT, dioxins are very persistent and "lipophilic," or fat soluble, so they can be easily stored in body fat. As a result they can accumulate up the food chain, so that species at the top of the food chain, such as man, have higher levels. The British Ministry of Agriculture has been monitoring levels in food and breast milk. They have confirmed that dioxins are present in low concentrations in all foods, especially fatty foods such as cow's milk.[12] Like PCBs and DDT, they are passed through the placenta and are excreted in breast milk, resulting in exposure of the fetus and nursing infant to this class of chemicals as well. Because different dioxins have different potencies, the measurements are weighted in terms of "toxic equivalents" or TEQ. In a recent study MAFF found the TEQ for human breast milk is 0.7 nanograms per kilogram. They found the infant receives its highest dose early on in nursing: estimated intakes from human milk are over four times higher at two months compared to ten months, due to an increased body weight in the baby and a move to a mixed diet. They report that the "mother's body burdens of dioxins will decrease during several months of breast feeding. The concentrations of dioxins in the mother's milk will also decrease over this period by about 12% a month." Not surprisingly, since women pass on their body burden of these persistent chemicals during breast-feeding, levels of dioxin from breast milk in mothers nursing their second child are as much as 20 to 30 percent lower.[13]

The ministry report concludes that although dioxin levels appear to be falling, "the estimated dietary intakes of breast-fed infants found in the current survey exceeded the recommended tolerable daily intake (TDI) and the Committee on Toxicity of Chemicals in Food, Consumer Products and Environment has considered at length the implications of this erosion of safety margins inherent in the TDI."[14] Tolerable daily intake for dioxin was recommended by a World Health Organization committee in 1990 at no more than 10 picograms per kilogram of body weight per day. The Ministry of Agriculture data indicate that a breast-fed infant at two months could be consuming ten times more than this, at 110 picograms TEQ per kilogram of body weight per day.[15] Despite this, the Ministry of Agriculture experts, along with many other health officials, still strongly recommend that breast-feeding should be encouraged because the balance of evidence suggests that the considerable benefits it confers in "immunological protection, nutrition and mother-infant bonding far outweigh the risks."[16]

Amazingly, recent studies investigating the mechanism of action of dioxin have revealed that it too can function as an "endocrine disrupter." Linda Birnbaum from the EPA's Health Effects Research Laboratory explains that among the first clues were the studies showing that dioxin could dramatically alter the effects of other chemicals in our bodies. "One of our first thoughts was that dioxin could in some way affect the thyroid gland," she recalls. "It is not like cyanide poisoning, where the animals die almost instantly. Dioxin causes delayed lethality. First there is wasting, just as in cancer; the organism may lose up to half its body weight. Now if someone has too much thyroid hormone you can also get very similar wasting syndrome." Furthermore dioxin could increase the incidence of cleft palate, a birth defect which can also arise with too much thyroid hormone. It was as if dioxin was somehow exaggerating the effects of thyroid hormone.

But there were signs that dioxin might also affect other hormones. Multigeneration studies showed dioxin affected reproduction and fertility. High doses suggested it could damage the testes and cause atrophy. Laboratory studies showed it could alter the levels of estrogen and progesterone.[17] Soon it was discovered it could modify concentrations of thyroid hormone in several tissues as well.[18] Then it was found to have effects on testosterone too.[19] The possibilities were endless. To add to the puzzle, it also had some anti-estrogenic action. Some research even suggested it might reduce the incidence of breast cancer, acting like the drug Tamoxifen, which is used in the treatment of breast cancer.

"It really wasn't clear-cut," explains Linda Birnbaum. "There were some effects that reminded us of too much of a hormone. Other effects appeared more like too little of a hormone. Sometimes it looked like too much thyroid hormone, sometimes too little, sometimes it was estrogenic and sometimes anti-estrogenic in its effects. Sometimes the hormonal effects were only in certain tissues in the body. And we began to think, wait a minute, dioxin messes things up entirely. It scrambles the hormone messages in several ways."

Dioxins, Gray explains, have been shown to exert their effects not by binding to the estrogen receptor, but to another crucial molecule, called the "Ah receptor." No one really knows what the Ah receptor does in the body. Although scientific understanding of the endocrine system is considerable, there are still vast gaps in our knowledge. Recently, scientists have identified several receptors, including the aryl hydrocarbon, or Ah, receptor, for which the normal chemical messenger or hormone has never been discovered. They are known as the "orphan receptors." Evidence is mounting that just as estrogen binds to the estrogen receptor and this enables it to activate the genetic material in the cell, in the same way dioxin binds to the "Ah receptor," which enables it to switch on or off several genes. Quite how dioxin disrupts the activity of so many hormones has remained elusive. It is likely, however, that once it has bound to the Ah receptor, it can then send out several signals which may alter the expression of many different hormones, argues Linda Birnbaum.[20]

"Dioxin is unlike anything we've seen in the literature," explains Gray. "It disrupts multiple endocrine systems. We're really quite concerned with some of the recent findings on dioxins. What is more, the fetus is especially primed to respond to dioxin. These Ah receptors can be found in the fetus and they have a different distribution to those in the adult. We think it plays a critical role in development and growth."

This is exactly what has been found. Recent studies have revealed that dioxin can exert its effects in the developing fetus at an incredibly tiny single dose. Much of the early work investigating the effect of dioxin on the reproductive system was carried out at high dose levels. Recently, Richard Peterson and his team at the University of Wisconsin tried something different. Their results caught everyone by surprise. In the Peterson study, pregnant rats were given one small dose of dioxin on the fifteenth day of the pregnancy. This is the day that is critical for male reproductive development, when the male reproductive tract begins to form. As adults, they found the male offspring exposed to dioxin had sperm counts that were up to 56 percent lower than in the controls.[21]

"Peterson's study showed the extremely dramatic effects at low doses," says Gray. Since dioxin risks were being reassessed by the Environmental Protection Agency at the time, this study caused quite a stir. Earl Gray and his team immediately set about reproducing the study and their research confirmed the Wisconsin result. Dioxin could have profound consequences on the developing embryo at one dose of less than a microgram per kilogram of the mother's body weight. This dose was too low to have any apparent toxic effects on the mothers. Yet the majority of the offspring showed permanent malformations in their reproductive tract that interfered with normal reproduction. It was a constellation of effects, including sharp reductions in sperm counts. Gray warned in his article describing these results that exposure to relatively low levels of toxic substances that disrupt the hormone system during critical periods in development can permanently alter reproduction and produce pseudohermaphroditism.[22]

Apart from reproductive effects, studies of dioxins have linked it with several human diseases. In addition to the disfiguring skin disease chloracne, which is the definite telltale sign of dioxin overexposure, it has also been linked to cancer. Studies by Pier Alberto Bertazzi of the University of Milan in a follow-up to the Seveso accident have shown higher than normal rates of several types of cancer. Liver cancer incidence was about three times higher than in the control population. There were also higher rates of leukemia, myeloma and soft tissue sarcoma.[23]

Dioxins have also been linked in animal studies to endometriosis. "This is an extremely painful condition that develops in women," explains Dr. Audrey Cummings from the EPA's Reproductive Toxicology Division in North Carolina. "Tissue from inside the uterus, called endometrium, finds its way outside the uterus to the abdominal cavity. In effect, uterine tissue grows where it shouldn't be, on the ovaries, bladder, bowel, or a number of other sites."[24] Over 10 percent of all women are estimated to be affected at some time in their lives. The first evidence that dioxin might promote this disease was found by chance. "More than ten years ago a primate study was set up for another reason: to find out if dioxin affected reproduction. Monkeys were fed very low doses of dioxin for a while. Several years later, the researchers found that many of the monkeys had developed endometriosis. The severity and incidence of the disease was greater in those monkeys that had received the highest dose."[25] In rodent studies at the EPA's labs, Cummings and her team confirmed that dioxin exposure promoted the development of this disease.[26] "I am very confident that endocrine disrupters, such as dioxin, could be associated with human diseases," she says. Her recent work is investigating how dioxins and other endocrine disrupters can affect pregnancy, especially disrupting the processes necessary for successful implantation of the egg.

Although scientists do not yet have a clear picture of how dioxin can have such a diversity of effects, studies such as these highlight a problem which Earl Gray and others have been arguing about for some years. Exposure to chemical pollutants may result in a myriad of subtle changes that can alter the balance of the hormone system through a variety of pathways. A single chemical, he points out, can alter the "endocrine milieu" through multiple mechanisms, to say nothing of the complex alterations induced by a mixture of compounds. What is more, he warns, "adult and fetal toxicology are not the same. If adults are insensitive to the effects of some of these compounds, that does not mean the fetus will be too."

In a recent publication Gray compiled a list of endocrine disrupters which highlights the possibility that many toxicants may alter thyroid and adrenal function as well.[27]

Endocrine Disrupters[28]

ENVIRONMENTAL ESTROGENS (ESTROGEN RECEPTOR MEDIATED)

Methoxychlor; some PCBs; β-isomer of lindane; o,p'-DDT; bisphenolic compounds, e.g. bisphenol A; octylphenol and nonylphenol/alkylphenol ethoxylates.

CHEMICALS THAT HINDER ESTROGEN ACTION

Dioxin (down regulates estrogen receptor at high doses); p,p'-DDT and DDE (increase degradation of estrogen in birds); endosulfan (inhibits vitellogenesis in fish).

ENVIRONMENTAL ANTI-ANDROGENS AND ANDROGENS (ANDROGEN RECEPTOR MEDIATED)

Vinclozolin (binds to the androgen receptor); p,p'-DDT.

ANTI-THYROID ENDOCRINE DISRUPTERS

PCBs; dioxin; PCDF; lead; many thiocarbamide- and sulfonamide-based pesticides; PBBs; phthalic acid esters; hexachlorobenzene.

ADRENAL ENDOCRINE DISRUPTERS

Vinclozolin and related fungicides; aniline dyes; carbon tetrachloride; chloroform; o,p'-DDT and DDE; dimethylbenzanthracene; methanol and ethanol; nitrogen oxides; PCBs and PBBs; fungicides of the ketoconazole class; toxaphene.

NB   The fact that a toxicant is on this list indicates that it may interact with the hormone system in some way, but does not indicate that humans are exposed to toxic dosage levels.

This list, Gray points out, is far from all encompassing. However, he argues that pesticides and toxic substances may attack multiple sites in the reproductive system and may alter thyroid and adrenal function as well. For example, certain PCBs, in addition to being estrogenic in their action, can also bind to the Ah receptor, "giving the exact same result as dioxin. . . . We know from animal studies that some PCBs can produce 'hypothyroidism,' where there is too little thyroid hormone." Studies have shown that too little thyroid hormone may have profound effects on human mental development, he explains, and is probably one of the leading causes of mental retardation. "Because recent studies have linked PCBs to behavior problems, this seems to me to be important. I think there is considerable cause for concern."[29]

"Nature is not designed by an engineer," says Professor Ana Soto. "We cannot get a clean classification for everything. For example, DDT is neurotoxic and it is also an endocrine disrupter with estrogenic effects. Yet its metabolites, or breakdown products, are anti-estrogenic and anti-androgenic and have immunosuppressive effects." In a recent publication summarizing the data on endocrine disrupters, Colborn, Soto and vom Saal warn: "Large numbers and large quantities of endocrine-disrupting chemicals have been released into the environment since the Second World War. Many of these chemicals can disturb development of the endocrine system and the organs that respond to endocrine signals in organisms indirectly exposed during prenatal and or early postnatal life. . . . Trans-generational exposure can result from the exposure of the mother to a chemical *at any time throughout her life* before producing offspring due to the persistence of endocrine-disrupting chemicals in body fat which is mobilized in . . . pregnancy and lactation. . . . Effects of exposure during development are permanent and irreversible."[30]

Reducing Exposures

Among the scientists there is general agreement that it is very difficult to take steps to reduce exposure, since many of these chemicals are so fundamental to our way of life. "I think it is very difficult to reduce exposure," explains John Sumpter, "because many of these chemicals are very widely used in a wide range of products. So, for example, if we decide not to buy milk that is in plastic containers because phthalates may be leaching into the milk in their containers and instead you buy it in glass bottles, then it is possible that the glass bottles were cleaned in detergents and these detergents might have contained estrogenic chemicals. Similarly, if you decide not to use canned vegetables because you are concerned about a possibility that bisphenol A might have leached into the can of vegetables and instead you switch to fresh vegetables, then these may contain estrogenic pesticides and herbicides. So, because there is a very wide range of chemicals that are estrogenic and widely used, it is extremely difficult, if not impossible at this stage, to give advice to anybody if they want to reduce their exposure."[31]

Nonetheless, many of the scientists interviewed in this book were asked if they had modified their lifestyle in the light of their findings. Despite the difficulties in weighing up the significance of different routes of exposure, most of them had made modifications. The following is a summary of key suggestions.

WATER

At present, whether using filtered, bottled or tap water, it is extremely difficult to know which reduces exposure most effectively. Several scientists, such as John Sumpter, have taken the step of filtering their water, although he points out that it is not proven that this reduces contamination. "We don't know which is the best kind of filter," says John Sumpter. "Most filters have a plastic container and we don't know what chemicals are used in the plastic. Then you have these charcoal filters which you have to regularly replace, and I don't know what is in the charcoal that might leach out. I wouldn't be staggered if we learned in ten years' time that I was adding more chemicals by filtering the water rather than getting rid of them. It is very difficult to win on this issue. Nonetheless, my wife and I chose quite a few years ago to filter our water and we will continue to do so."

Ana Soto has chosen to use bottled mineral waters but she is still concerned about these. "We use spring water, but spring water comes in containers here that look suspiciously as though they have been made of polycarbonate. . . . I wonder when we are going to know. It would be so nice if we knew at least which containers were made of which substances, whether they are leaching bisphenol A or phthalates." At present this information is not available.

Despite all these problems they urge for a balanced perspective. "Life expectancy has gone up immeasurably this century," says Sumpter, "and a fair bit of that is because of improvements in water and food quality. There used to be many water-borne diseases, such as cholera and hepatitis, and now these are gone in the West. Overall, I don't doubt that improvements in water quality this century have saved millions of lives. Somehow, we have to strike a balance between concerns."

AVOIDING FATS

There is a considerable body of evidence to suggest that reducing consumption of fats, especially animal fats, may help to reduce exposure. A great many of the chemicals which are causing most concern are the "lipophilic," or fat-loving, compounds which bioaccumulate in fatty tissue. As shown earlier, DDT and other pesticides, PCBs, and dioxins, all accumulate in our own fat stores and may remain for months or years. "Many of these chemicals travel through the food web in *fat*," write Theo Colborn and her colleagues Dumanoski and Myers, "and become more concentrated as they move upward to the top predators such as polar bears and humans. . . . Eating less animal fat, found in foods such as butter, cheese, lamb, beef, and other meats will greatly reduce exposure to hormone-disrupting chemicals."[32]

Because these chemicals accumulate in fatty tissue they have been shown to accumulate in particularly high levels in breast milk and breast milk fat. This raises the vexed question of feeding the newborn, which is discussed in detail in Chapter 8. A number of studies have highlighted the problem that many of these chemicals are transferred to the infant during nursing in high levels close to, or even exceeding, recommended levels in the case of dioxins and PCBs.[33] Although cow's milk, from which most formula milk is made, carries a much lower burden of these compounds, other chemicals have been found in formula, such as butylbenzyl phthalate. Most experts continue to recommend breast-feeding. Colborn and her colleagues highlight the "pressing need for more research" on this issue, since, they argue, the transfer of a woman's chemical load built up over her lifetime into her newborn is undesirable. "It is critical," they write, "that we . . . make choices that reduce this chemical legacy which is being passed on from one generation to the next. . . . Children have a right to be born chemical-free."[34]

Apart from taking steps to reduce exposure to contaminated foodstuffs, there is also the question of avoiding contaminated packaging. Recent research outlined in Chapter 8 shows that some of the estrogenic chemicals used in some plastics can leach from packaging into fatty foods. These can include dairy products such as cheeses and butter, and other fatty foods such as chocolates, pies and potato chips. Most of the scientists interviewed are avoiding foods wrapped in plastic where possible, or removing it from the packaging as soon as possible. Not all plastics leach chemicals, but at present, in the absence of labeling describing the constituents of the plastic, it is very difficult for the consumer to have any way of knowing.

"We never use plastic-wrapped dinners that you heat in the oven or microwave," says Professor Sumpter. "I am concerned about these. People think of molecules as though they are fixed in one place. But they are not; they move about, whether in solids or liquids, and the higher you heat them the faster they move. So I would anticipate that if you heat something up in plastic it is more likely that the transfer of molecules into the food will be greater as the temperature goes up."

"Certainly, I'm not using plastic as much as I did before," says Soto. "I'm not heating anything in a plastic container. Also, if possible, when I'm shopping, I put my vegetables in a paper bag and then in a larger plastic bag to avoid direct contact."

The marketing of food has changed enormously in the last twenty years, with considerable emphasis on packaging which is convenient for transport and storage, and gives appealing presentation. Nonetheless, in the light of concerns over the leaching of bisphenol A and phthalates from packaging, there are growing concerns that this may need to be tested and labeled, so consumers know exactly what they are eating.

AVOIDING PESTICIDES

Pesticide residues can be detected in many of the fruits and vegetables we consume and it is very difficult to know how to reduce exposure to this. "We've always washed our fruit and vegetables. But if you ask me what is the evidence that washing fruit and vegetables really reduces exposure, I can't produce it," says Sumpter despairingly. "We just don't know." Some of the chemicals used are not water-soluble. Others may be "systemic" and can penetrate inside the product.

Professor Skakkebaek has taken this issue into his own hands. "To some extent I have altered my lifestyle. I don't use pesticides anymore in my garden—it's an organic garden—and also I'm eating organic butter which you can buy here."

Colborn points out that many pesticides currently in use have never been tested for hormone-disrupting activity by the Environmental Protection Agency. She cautions against the casual use of pesticides in gardens and points out that studies have found higher rates of cancer in pets such as dogs in households where pesticides are used.[35]

Britain has one of the smallest percentages of land under organic production in the European Union, despite a huge increase in demand. Some European countries, such as Austria, now have as much as 11 percent of the land under cultivation with no chemical pesticides or fertilizers. In America, demand for a "clean food diet" has increased significantly, fueled by recent pesticide and BSE scares. The U.S. Food Marketing Institute shows that 42 percent of mainstream supermarkets carry organic produce and a quarter of all shoppers are buying organic food once a week. If demand continues at the same rate, it is estimated that organic food will increase to 20 percent of all food sales by the turn of the century in America.

PHYTO-ESTROGENS

Dr. Richard Sharpe is most concerned about the increasing use of phyto-estrogens in baby foods until we have further data. "The evidence for soya having beneficial effects in adults looks quite strong . . . but it does contain a lot of potent things, and I'm not happy for children who are still growing to be eating soya. I don't think it is right that we are giving them food which contains hormones without knowing exactly what effects it has. Phyto-estrogens are produced in nature to actually exert reproductive effects on animals and predators, at least that is the thinking. . . . I'm not saying that it necessarily has an adverse effect, but there are plenty of animal studies which suggest it might. Alternatively, it might even have beneficial effects. Nonetheless, I am concerned about our blind use of these substances in baby foods." When buying for his young family of four, he will try to avoid products with added soya for the children. Other scientists also expressed concerns about soya in the diet for young children, including Professor vom Saal and Dr. Risto Santti.

One of the overriding concerns expressed by many involved in the research was the difficulty of giving specific advice, since so little is known about routes of exposure. "I have not been able to alter my lifestyle very much," explains Skakkebaek, "because my problem is that I do not know how to alter it. I don't know what should be done. If we are going to be very rational, these chemicals are so widespread we're going to have to wait for more data."

Sumpter, too, highlights the gaps in our knowledge. To date, for example, there has been very little research on cosmetics and health-care products. "Many face creams will almost certainly have surfactants in them to help them spread," he says. "Some shampoos also have surfactants—that's how they work." There are other unknowns too. "To my knowledge, no one has begun to look in detail at exhaust fumes to check out what hormone-disrupting chemicals may be there. Dioxins have been identified and it would not be surprising if there were related compounds. Some petrol has surfactants in it. These are all complex mixtures which need to be tested. . . . There is a good argument for carrying out tests on *products,* rather than individual chemicals, as at present."

For Ana Soto it is almost like being thrown out of the Garden of Eden. "It has changed things for me. I've lost the innocence I had before, thinking I could drink what I liked and it wouldn't do any harm. Now that I know all these things, I really wonder, do I have to drink as much water as I'm drinking, or eat what I am eating? If you are doing something that is not strictly necessary, you are just accumulating all this stuff in your body. That has changed me. Now I think that if you don't need it, why would you? Now I know that every time you eat and drink you could be adding to the chemical burden in your system. It's as though all these products that we've synthesized are out there and getting back to us, no matter what we do."

CHAPTER TEN

"Playing the Trump Card of Uncertainty"

Some scientific papers, once published, seem destined for obscurity, yellowed manuscripts of learning relegated to some dim and dusty archive. Not so Professor Niels Skakkebaek's paper in 1992 suggesting there had been a 50 percent drop in sperm counts in the last fifty years.[1] Prompted by concerns arising from Skakkebaek's own observations in the fertility clinic at Copenhagen University Hospital, he and his team had analyzed more than sixty studies on sperm counts carried out over the last fifty years. These historical studies were from many countries worldwide, and revealed that men born in the 1930s and 1940s had much higher sperm counts than men born in the 1960s and 1970s. Although it wasn't intended as such, it would be hard to conceive of a bolder challenge to comfortable medical complacency, which had always presumed, despite a number of reports to the contrary, that there was not much to worry about. In next to no time, a debate took off in the pages of learned scientific journals.

Dr. Stewart Irvine at the MRC Reproductive Biology Unit in Edinburgh recalls the response to Skakkebaek's paper: "There was a huge controversy and a lot of criticism of the Danish paper has been published. It's interesting that at least some of the criticism has emerged from *industry*. There were even some papers that clearly had industrial authors on them."[2]

One study in 1995 on the subject of declining sperm counts was undertaken by Geary Olsen and his team of the Dow Chemical Company Epidemiology Department, with Charles Ross of the Corporate Medical Department, Shell Oil Company, and Larry

Lipshultz of the Baylor College of Medicine. They reanalyzed the sixty historical studies in the Danish paper, using different statistical techniques, and arrived at the conclusion that, in the last twenty years, sperm counts may even be *slightly increasing*. The authors found that Skakkebaek's study was flawed, since they believed the type of statistical analysis used by the Danish team, known as "linear regression," was inappropriate for the data. When they used more complex statistical techniques, the "quadratic" or "spline" models, they were not persuaded that there had been a continuous downward trend in sperm counts since 1940. Although they did agree that all statistical models indicated a reduction in sperm counts over time, they argued that a "stair step" model, suggesting that there may have been one "Niagara Falls" drop in the 1960s, with sperm counts constant since then, was a much more plausible interpretation of the data.[3]

They were not the only critics. Discussions centered on how the historical studies were selected, whether the men studied were representative of the population as a whole, were the statistical analyses weighted correctly, was the data skewed, and was linear regression the most appropriate model. In short, every aspect of the Danish paper was pored over in minute detail. But as it turned out, Professor Niels Keiding from the statistical research unit at Copenhagen had already tested other statistical techniques before they published their findings. While acknowledging that they would, of course, wish for better historical data, Keiding pointed out to his critics: "On the basis of the admittedly fragile instrument of meta-analysis, the evidence cannot lead us to a conclusion other than that sperm concentrations now are lower than they were in the 1940s and 1950s. . . . We do not believe that further statistical analysis will make any difference to the main message."[4]

In fact, Skakkebaek and Keiding's paper was not the first review of the data to arrive at the conclusion that sperm counts were declining. In 1980, James and his team, reviewing twenty-nine studies on over 8,000 men, also found a marked decline, from a sperm concentration of over 130 million in 1930 to less than 80 million in 1980. But such historical precedents seemed to be overlooked in what was becoming an increasingly heated debate.[5] "People were just disbelieving," recalls Dr. Ewa Rajpert-De Meyts from Copen-hagen University Hospital. "Some tried to poke holes in the sta-tistical analysis to try to show their disbelief. Others pointed out that sperm counts are a parameter that is extremely variable from day to day, depending on sexual habits, drinking, smoking, and so on. . . ."[6]

Sharpe recalls one meeting organized by industry. "It felt like a 'kick us into touch' meeting," he says, attended by a great many people who were sympathetic to industry's point of view and who didn't agree with the Danish study. "There are people who just say, I don't believe it! To me that is a completely illogical argument. You can't stand up and simply say, I just don't believe sperm counts are falling. My reaction would be, what about the evidence? They say, I just don't believe it. I think you lose all scientific credibility to react in this way. It is like somebody publishing an experiment in a refereed journal and people saying, 'Look, I'm sorry, I just don't believe it. It's rubbish!' If we operate on that level, then we'll never get anywhere. We need to form opinions based on *evidence,* not on prejudice, emotion or commercial concerns." He recalls the atmosphere of the meeting. "It was like a broadside trying to de-stroy the Danish paper. . . . About halfway through the meeting, the person next to me leaned over and whispered, 'I think you've been set up!' This person actually made the point (I was going to make it, but he beat me to it, which is probably just as well) when he said, 'Hang on a minute. Aren't we all overreacting a bit here? I've got a copy of Skakkebaek's paper and it didn't make any of the claims that you're saying. It actually says that the data suggests there might be a problem with sperm counts, but because of the limitations of the data, we need further work.' "[7]

But while statisticians scrutinized the mathematics, male fertility experts hurried to check their own sperm count data. In Paris, Bel-gium, Edinburgh, New York, London and Helsinki, new studies were begun in an attempt to determine whether the decline was real. Meanwhile the "estrogen hypothesis" took on a life of its own. This had become the fashionable thesis. Here at last was an important and testable model of environmental hazard which had implications for the public and policy-makers alike. From the corridors of Whitehall, to the labs of far-flung universities, pressure to establish whether or not it was true was mounting. In Brussels, Washington and Whitehall, committees were formed, experts were summoned, reports were prepared, memos were drafted. No minister, senator or MEP would be caught without an opinion. But no amount of reanalysis of old data would settle the issue. What was needed was some new data. This was soon forthcoming and, while a great deal of it confirmed the Danish findings, it was also clear there were some inexplicable but marked regional variations.

Paris, France

Among the skeptics when Skakkebaek and his team published their paper on falling sperm counts in 1992 was the French fertility expert Pierre Jouannet, Director of the Centre d'Etude et de Conservation des Oeufs et du Sperme, Hôpital Cochin, Paris. Jouannet and his team were convinced that there had been no change in sperm quality over the years, but felt it was important to assess Skakkebaek's report.

The Paris sperm bank, created in 1973, at the University Hospital provided a unique record of male reproductive health. They realized they could easily put Skakkebaek's ideas to the test. They had been using exactly the same testing procedures on a constant population: healthy young men who were donating sperm. They included over 1,300 men for their study, all unpaid volunteers living in the Paris area who had previously fathered at least one child.

It took some months to analyze all the data. Numerous measurements were made: the volume of ejaculate, the sperm concentration, the motility, or movement, of the sperm, and the abnormalities in shape. The statistical analysis was equally complicated, taking account of the year of donation, the age of the donor, the year of birth, the duration of sexual abstinence, and many other factors.

To their considerable surprise, they found there were indeed dramatic changes in sperm counts. The average sperm count of men tested in 1973 was 89 million per milliliter. This declined to

60 million per milliliter in 1992. This was an average decline in sperm counts of more than 2 percent per year, which is extraordinarily fast for a physiological change. They also found a decline in motility and percentages of normal sperm.

They were so concerned by this result that, just like the Danish team, they considered every possible pitfall in their study. They wondered if social or lifestyle changes could affect the result. If, for instance, men were having sex more frequently now than before, this would bias the result in favor of lower sperm counts. But every line of inquiry seemed to confirm their findings. Sex was no more frequent now than in previous generations: the average number of sex acts remaining constant, at 9.3 times per month. In fact, when they took a subset of men, matched for age and duration of abstinence they found an even greater drop in sperm counts: from 101 million per milliliter in 1973 to 50 million per milliliter in 1992. The rate of decline in sperm counts in this group was even faster, at 3.7 percent per year. They even analyzed the number of technicians involved in the counting to see if this could have involved any changes that could skew the results. It was soon apparent that during the twenty years of study there were few changes in the staff: "All technicians had the same training . . . and there was no new equipment introduced."[8] There seemed to be nothing that could account for the changes, other than that a real drop had occurred.

It took some time for these results to be published in the prestigious *New England Journal of Medicine* in February 1995. In the article, Jouannet concluded: "There has been a true decline in the quality of semen during the past 20 years. . . . The fact that not only the concentration of sperm but also the percentage of normal sperm declined indicates a qualitative impairment of spermatogenesis. . . . This decline is unexplained."[9] In an interview with BBC *Horizon* researcher Diana Hill, he said: "The situation is really urgent. We may be seeing now the effects from 30 to 40 years ago and we need to act fast because it could get worse. We cannot sit back and wait. We are not saying we have proved the decline in male fertility, but that we think it is probable."

But, perhaps somewhat unusually, when Jouannet's paper was

eventually published in the *New England Journal of Medicine,* a critique was included within the same edition. Richard Sherins from the Genetics and IVF Institute questioned whether the men in Jouannet's study were representative of the whole population, since they were men serving as sperm donors, attending infertility clinics or about to undergo vasectomy, groups which he regarded as unrepresentative. He pointed out that despite Jouannet's attempts to control for variables affecting sperm concentrations, such as age and duration of abstinence, there can be variations in samples even from one man and therefore relying on a single sample can be "hazardous."[10]

But despite the difficulties of measuring sperm counts, two years later French data from the *in vitro* fertilization register confirmed the worrying trend identified by Jouannet. De Mouzon and his team reviewed the results of over 7,000 normal healthy men whose partners were undergoing IVF for tubal disease. This is the largest group of men that has been studied. They, too, found a significant decline in sperm count depending on the year of birth. The average sperm count for men born before 1939 was 92 million per milliliter, whereas men born after 1965 had rather lower sperm counts of 77 million on average.[11]

Flanders, Belgium

It wasn't long before two further European studies provided more evidence for impaired sperm quality. The first of these was from Professor Frank Comhaire, from Flanders in Belgium. He too had been present when Skakkebaek originally presented his data in 1992 at the World Health Organization conference. Professor Comhaire and his team had established a sperm bank at the University Hospital in Ghent, nearly twenty years previously in 1977. Like the French team they realized they could easily check whether their data tallied with Skakkebaek's findings. They studied data from over 400 healthy young men who were wishing to donate sperm, comparing samples from men taken in 1977–80 with those taken in 1990–95. Although they found sperm concentration had declined by 12 million per milliliter during this time, sperm ejaculate volume had increased slightly, so on average there was no change in sperm count. However, they found other worrying alterations to sperm production, which they argued "are so important as to threaten male fertilizing potential."[12]

"Two important measures of sperm quality had deteriorated over time," explains Professor Comhaire. "Firstly, motility: to reach the egg, the sperm needs to move at a certain speed. Grade A sperm are those that can move at 8 centimeters per hour and can therefore reach the egg in good time." They found the number of grade A sperm, with rapid movement, had declined substantially from 52 to 31 percent.[13]

"Secondly, once the sperm has made it to the egg, what matters is its shape. Does it have a normal morphology?" This was checked under the microscope on a stained and dried sample and was defined as the absence of any obvious clear-cut abnormality. The average proportion of sperm with a normal shape had declined from 39 percent in the late seventies to 26 percent in the nineties.[14]

"Because there were adverse changes in both movement and shape, when these are measured in different ways, this supports the view that what we observe really is a deterioration in sperm quality," concludes Comhaire. Asked if he thought the changes were important, he replied, "We don't think it, *we know it*! The number of severely subfertile men has increased. This is defined as those with a very low count of 4 million sperm per milliliter or below, or only 3 percent rapid motility, or 6 percent normal forms. There was a fivefold increase in this group from 1.6 percent of the population before 1980 to 9 percent of the population since 1990. That is a fivefold increase in men who will have a serious infertility problem and that is pretty worrying!"[15]

London, England

In London, Dr. Jean Ginsberg in 1994 reported similar adverse changes. She reviewed data from healthy normal men whose partners were receiving fertility treatment at the Royal Free Hospital in London. She compared changes in sperm quality and concentration between men living in central London, within the Thames Water supply area, and those living outside it. Remarkably, there were differences between these groups. Men living in central London, within the Thames Water area, showed a significant decrease in sperm concentration, from 105 million per milliliter in 1978–83 to 76 million per milliliter in 1984–89. Men living outside this area did not show any such dramatic declines in sperm concentration. However, all men studied, irrespective of where they lived, showed other adverse changes, reminiscent of Comhaire's study, most noticeably a substantial increase in the number of abnormally shaped sperm. She was intrigued that there could be differences in sperm quality depending on where the men lived. However "despite exhaustive enquiries we were unable to ascertain whether there had been any change in Thames Water testing of water supplies at source and delivery after 1983."[16] Besides, she acknowledges, it may not be the water; there could be other environmental factors involved in central London. Overall, she concluded that there has to be some general environmental factor involved since all men showed a significant increase in the number of abnormal sperm.[17]

Edinburgh, Scotland

The most recent British study also brought worrying news. In Edinburgh, at the MRC Reproductive Biology Unit, like Jouannet and Comhaire, Dr. Stewart Irvine was also full of doubts when he first heard of the Danish study reporting a 50 percent drop in fifty years. By chance, his team, too, had been collecting data on sperm quality for research on sperm cell biology. They realized this data too might contain a hidden message.

Analyzing their records on nearly 600 men, they began to check out the correlation between the year of birth and sperm count. These preliminary results were striking. They found an average sperm count of 98 million per milliliter for donors born before 1959 and an average of only 78 million per milliliter for those born after 1969. At first glance this trend was in strong support of Skakkebaek's findings and this prompted Dr. Irvine and his team to embark on much more detailed study.

They divided their data into four sets according to the year of birth: men born in the 1950s, the early 1960s, the late 1960s and the early 1970s. The youngest group of men they had in the study were eighteen or nineteen, the oldest were in their early forties. Conscious of the criticisms that had greeted earlier studies, Dr. Irvine took care to avoid any pitfalls that might bias the result. "All the samples were analyzed in one laboratory by one group of people using the same methods throughout the period of study," he explains. "The men were recruited simply because they had volunteered to help us with research; they were not recruited because of any particular fertility status or other such criteria."[18]

"One would expect semen quality in older men to be less good than in younger men. The perceived wisdom has always held that semen quality deteriorates with age," he says. "In fact, what we found was the reverse of that. There was a much more substantial decline than I would have anticipated over time: the younger men, born in the 1970s, had sperm counts that were some 30 percent lower than men born in the 1950s. . . . What is more, there was not just a decline in sperm count but also in sperm quality, in the number of active moving sperm.[19]

"In terms of numbers, men born in the 1950s had an average sperm count of about 100 million per milliliter," says Dr. Irvine, "whereas men born in the early part of the 1970s had an average sperm count of about 75 million. Now clearly *if* this decline is continuing and it carries on down, then men who are being born now in the early part of the 1990s are going to grow up and have sperm counts of about 50 million. That's worrying, because in an infertility clinic we tell a man that his fertility is likely to be compromised when his sperm count is below 20 million. . . . All the new data that has been published has tended to point in the same direction, suggesting that we do have a problem to deal with. That makes me stop and be worried frankly—this is a substantial change that appears to be continuing."[20]

But, once again, this study too had its critics. It was beginning to look as though it was impossible to design the perfect study on sperm counts that would satisfy the skeptics once and for all. Writing in the *British Medical Journal* about Irvine's study, Professor Stephen Farrow from Colindale Hospital, London, explains: "It seems plausible that the older the donor recruit, the more likely he is to come from the antenatal group (that is, to be of known fertility); and the younger the donor recruit, the more likely he is to come from the undergraduate group (that is, to be of known fertility). . . . This is an exceptionally skewed population and not appropriate as a basis for drawing general conclusions."[21]

While it may not seem surprising that young men are much more likely to be of unknown fertility status than older men, it was becoming clear that sperm quality was a surprisingly difficult topic to study. Apart from the statistical models used, numerous other factors that might affect sperm counts had to be assessed: age, occupation, fertility status, duration of abstinence, drinking and smoking, lifestyle, and so on. To find a large group of men where one could control for all the factors that may have a bearing on sperm counts was beginning to look like an impossible task. In an emotional debate, was it possible that too much was being asked of science to provide "certainty" where it could only provide guidance to trends?

"I don't think taking each little bit of evidence in turn," says Irvine, "and saying okay we can push that brick out of the wall because it had that little problem and we can take another brick out of the wall because it had another little problem is going to get us anywhere. *All* of the data that's been published has problems, and all of the criticisms can be countered in various ways. That is how science works: building up a mass of data, some of it poor quality, some of it good quality. . . . I'm becoming increasingly skeptical that we'll never definitively answer the question of whether or not there has been a decline," he says, "simply because the study that we'd have to design would be so complex it would be impossible to arrive at a robust conclusion because we can't take account of all the confounding factors. . . . *I think we have to look at the global totality of the data and ask, on the basis of this large mass of data, if we can now be comfortable that all is well. My feeling at the moment is uncomfortable, that we have a problem, and that we need to collect more data to address this.*"[22]

In support of this view, in addition to the new data showing a decline in sperm quality and count from Paris, Flanders and Edinburgh, recent studies in many countries report low sperm counts: in Nigeria, 64 million; in Pakistan, 79 million; in Germany, 78 million; in Hong Kong, 62 million.[23]

Interviewed for *Horizon* shortly after the publication of this data, Richard Sharpe summarized the concerns: "What all the recent studies have found is that there has indeed been a fall in sperm counts in these centers, Belgium, Paris and Edinburgh. This data is rather convincing because there are no substantial criticisms that you can apply to it. The upshot is that sperm counts have dropped by something like 2 percent a year in those centers which have made these measurements. Over a twenty-year period that amounts to a 40 percent drop, which is a considerable drop in sperm counts. I think it approaches the point where we would expect that it would have some impact on male fertility."

Asked if we were gambling with the future of the species, he replied: "If the fall in sperm counts were to continue and to continue at its present rate, then it's not too far in the future before we would be facing widespread male infertility. I think we can have relatively little doubt about that. . . . It is obviously worrying when you don't know the cause of something and when we don't know if it has stopped. It might be a progressive change, so I think that we have to do something now, before we get into really dire straits. We need to know the extent of the problem, how many countries, how many cities, and whether or not the problem is getting worse."[24]

Regional Differences

But there are some vocal dissenters from this view. According to Harry Fisch, Associate Professor of Urology at Columbia University, in America, all is well. At a meeting of the American Urological Society in Florida, he announced in May 1996 that American sperm counts were not in decline at all, and if anything there may be evidence of a slight increase in the last twenty-five years. His team studied data from over 1,200 men, who donated sperm before having a vasectomy in the years 1970–94 at three of the oldest and largest sperm banks in America. They also reported very marked regional variations. The average count in New York was 131 million per milliliter, the same as reported in earlier studies; in Minnesota it was 100 million, and in California, the lowest at 72 million. Since much of the older (pre-1970) data in Skakkebaek's original study was from New York, where sperm counts are high, this, Professor Fisch points out, could have skewed the Danish analysis. He concludes: "Prior studies suggesting decreases in the semen quality over the last few decades have been fraught with selection, methodological and geographic biases. By attempting to minimize these biases we found no such decline in semen quality."[25]

Apart from these regional differences, he raises another issue. "We've also found that sperm counts fluctuate from year to year. In fact the fluctuations are so dramatic you need a very long period of time to make a determination of a trend. . . . For instance, had we initiated our study, not in 1970 or 1971, but in 1977, 1980 or 1984, we also would have found a *decline* in sperm counts since these years had particularly high sperm counts, whereas between 1971 and 1976 sperm counts were relatively low. . . . The take-home message is that yearly fluctuations are so dramatic they limit our ability to make conclusions over trends. . . . Even though we found a slight increase over the twenty-five-year period, that is influenced by the year in which we started our study."[26]

Professor Stephen Safe from Texas A&M University also cites this study to question the hypothesis that sperm counts are declining. "I don't think you can do meta-analysis on worldwide studies when there are such regional differences," he explains. Pointing out that much of the early data for the 1930s and 1940s in Skakkebaek's study was from New York, where sperm counts are high, he adds. "So the whole thing has collapsed."[27]

But many specialists in male fertility are not convinced by Fisch's data. "His patients are men presenting for vasectomy, so they are a very selected group of older men, who are likely to be highly fertile," says Richard Sharpe. "In fact, it was already shown in Pierre Jouannet's studies that men presenting for vasectomy have high sperm counts, and so the French study specifically didn't include them because they are likely to be unrepresentative."[28]

"This would also tend to give a certain age bias," adds Dr. Stewart Irvine. "You don't usually do vasectomies on men in their twenties, so that would tend to obscure any secular trend showing a decline in sperm counts. . . . Furthermore, the three separate centers did not use the same methods to count sperm. They did not even use the same methods consistently within one center during the whole of the study. . . . So I don't think the paper was, as presented in some quarters, near definitive evidence: end of discussion! What was markedly lacking, given the striking regional variation they saw, was any explanation of why that might be. This was the most dramatic aspect to the paper. It seemed to ignore its own most important finding, namely, why sperm counts are twice as high in New York as in California." Professor Niels Skakkebaek agrees: "If this is true, this in itself would appear to indicate an environmental effect, would it not?"

"All I'm saying is, wait a minute, put a halt on this, we just don't have enough data to make these conclusions," says Harry Fisch. Highlighting regional and yearly fluctuations, combined with the fact that no study has ever investigated the general population, he concludes: "It may be going down, or it may be going up, but we don't have enough data to make these determinations!"

There is other evidence for marked regional variations in sperm count. However, with the exception of Fisch's data from New York, the regional differences are greatest between rural and urban populations. The most striking contrast is between Denmark and neighboring Finland, which is much more rural. Data from Finland shows sperm counts here averaging 133 million, the highest reported in any country since 1956, literally twice as high as Denmark. In addition to this, rates of testicular cancer are also low, a quarter of those of Denmark and other reproductive abnormalities are lower. The authors point out that Finland, which has been slow to industrialize and is still predominantly rural, could be avoiding the environmental factors which may be depressing sperm counts elsewhere.[29] In Toulouse in France, a study by Bujan and his team pointed to a similar trend. In the last twenty years in Toulouse they found no evidence of changes in sperm quality with time, unlike Jouannet's data from Paris. Noting that pollution levels and population density are all greater in Paris than in Toulouse, they argue this strengthens the case for further environmental studies.[30] A European study is now under way to investigate the reason for these pronounced differences.

While, for many, the regional differences make an environmental effect more likely, Stephen Safe himself argues the relationship with chemicals becomes less strong. "Basically we don't have chemical gradients between us . . . we are all contaminated by about the same amount. Now I suppose there may be regional water differences where there are environmental estrogens, but I seriously doubt it. It is possible. But I don't think New York, Minnesota and California, for instance, can have chemical gradient differences. . . . I don't think a chemical is involved and I don't think it's right, whenever there is a problem, to blame a pesticide or a chemical. Now in some cases this is correct, but I think this *chemophobia* has gone too far."[31]

But, for many, this conclusion is premature. A recent Danish study showed that men who had lower exposure to certain chemicals through their diet had higher sperm counts. Sperm counts from members of the Danish Organic Food Associations, who had a diet consisting of at least 25 percent organic produce, which is made without use of pesticides or chemical fertilizers, were compared to a control group who worked for an airline company. The average sperm counts of men eating organically produced food were 43 percent higher: 99 million compared with 69 million in the control group.[32] In Spain, another study also highlighted a possible link between male reproductive disorders and exposure to pesticides. A team from Granada University found the incidence of undescended testes in boys was significantly higher in those agricultural regions where pesticide use was greatest.[33] Furthermore, in reply to Safe, scientists point out that if, as suggested, the mechanism is arising from fetal susceptibilities to chemicals in the womb, measuring chemical gradients between adults could be meaningless. Many other factors could be involved, including the timing of exposure to suspect chemicals, individual differences in metabolism and immunity, and so on, making this an extremely complex issue to assess.

So after four years of debate, quite apart from speculation over the cause, can science even answer the question of whether or not there has been a decline in sperm counts? Recently, the World Health Organization commissioned an independent scientist, Dr. Elsebeth Lynge from the Danish Cancer Society, to review the data. She noted that, despite the debate, "all . . . models used to analyze the original Danish study pointed to a change from a higher level in the 1940s to a lower level in the 1980s, and a decline was also seen for studies originating from the United States. *It is my personal opinion that there is sufficient evidence for a decline to warrant concern.*"[34] Like many of the scientists involved in the discussions, she highlights the need to take the broader picture into account, changes in incidence of such reproductive abnormalities as hypospadias, testicular non descent and testicular cancer, all of which, as pointed out by Sharpe and Skakkebaek, show an increase.

"I certainly believe that we have a problem with male reproductive health," concludes Skakkebaek. "It is quite clear that the best evidence that we have is from testicular cancer. But it is also evident that all the trends that are available go in the same direction, namely, poor semen quality. I don't think there is any new evidence against these concerns."[35]

"Do we have good evidence, arguments, preferably data, that would allow us to explain away all the reports of falling sperm counts?" asks Sharpe. "The answer is that we don't, especially when you consider as well the data on undescended testes, hypospadias and testicular cancer."

It was studies on testicular cancer which first prompted their interest. Several new studies have confirmed the worrying trends they found. Since this is data from cancer registries, it is seen as very reliable: there is little chance of testicular cancer being misdiagnosed. For instance, a Swedish team, drawing on cancer registries from over nine different countries, studied over 34,000 cases and found that the incidence of testicular cancer increased *rapidly* in all the countries over a forty-year period. The average increase in Nordic countries was approaching 3 percent per year, and in

Germany and Poland, 5 percent per year. In fact the increase was so marked, they argue that if the trends were to continue at the same rate, the overall occurrence of testicular cancer will double every 15 to 25 years. The authors conclude that there are good reasons for believing that some kind of environmental factors must be involved and, what is more, that this exposure must start early in life, or even prenatally, since testicular cancer principally affects young men: "More probably the testis is particularly susceptible to environmental carcinogens at young ages." In agreement with Sharpe and Skakkebaek, they argue that investigating exposure to sources of estrogens in order to understand the increase in this cancer was "highly justified" by the data.[36] Two other recent studies have confirmed this trend, one from America[37] and one from Norway, which found incidence had tripled in the last forty years.[38]

"At present testicular cancer affects almost 1 percent of all men in Denmark," says Skakkebaek. "I think 1 percent is bad enough. This is a very significant figure. What is more, the data shows this may be due to events in fetal life, supporting the hypothesis that it may be caused early in life. . . . We cannot *prove* that environmental estrogens are the cause. We have to admit that semen quality is a difficult topic to study. However, the most cautious conclusion that we can draw at the moment is that there is certainly a problem with male reproductive health and it is getting worse. That is the most cautious!" Asked if the controversy over sperm counts had harmed his reputation, he just laughed. "How would I know? I think it's good to have a debate. I think that's fine. I don't think it's going to harm me very much!"

But there is one drawback to a lingering debate which is expressed in a recent editorial in the scientific journal *Nature*. The problem with scientific uncertainty is that it may leave room for vested interests to promote a biased perspective. While acknowledging that the statistics of sperm count decline are controversial, they point out that hormone-disrupting chemicals affecting the fetal testes are prime suspects. The authors argue: "In assessing the potential impact of such pollutants, *the current limitations of science are all too apparent. . . . It can only accomplish so much.*" Whenever "issues of pollution have become politicized, uncertainties within the science provide a notorious opportunity for conflicting parties in any environmental dispute."[39]

For in dealing with complex environmental issues, science can highlight trends, but cannot provide black and white answers. This will always leave room for lobbyists, industrialists, manufacturers—anyone with a stake in the argument—to "play the trump card of uncertainty." Because, for all the sophistication of modern science, and despite the wealth of data going in the same direction, it isn't possible to give an answer now, which is beyond argument, to the question of sperm count decline. This could be the ultimate scientific cul-de-sac. For the irony of the situation is that if the "trump card of uncertainty" is played for too long, and too loud by those with a vested interest, it may be too late before we find out. "In the long-term game called evolution," warns Professor Carlos Sonnenschein, "there are unpredictable winners and losers. It would not be too clever for humans to inadvertently load the dice against their own chances."[40]

CHAPTER ELEVEN

The Human Price

If the increase in environmental estrogens is the reason for the increase in breast cancer, then this is killing more people per year than Hitler did during the whole war, and that is completely outrageous[1]

DR. HENRIK LEFFERS, *molecular biologist,*
Copenhagen University Hospital

It doesn't appear at the moment that there is a relationship between these compounds and breast cancer.[2]

PROFESSOR STEPHEN SAFE, *toxicologist*
Texax A&M University

The scientific debate over a possible link between hormone-mimicking chemicals and breast cancer is difficult to resolve because of the complexity of the disease. Although the incidence of breast cancer has increased dramatically in the last fifty years, the causes are still poorly understood. "There must be more than 10,000 scientific papers *at least,* investigating the causes of breast cancer," says Leffers. "Yet still no one knows the cascade of events that are thought to be involved."

"We do know that between 5 to 10 percent of breast cancers, at most, have a genetic element and there is a search on for breast cancer susceptibility genes," explains Dr. Philippa Darbre, who specializes in cell biology at the University of Reading. "But for 90 percent of the breast cancers, we simply don't know the cause, although there must be some environmental influence." Environmental influences are suspected, she explains, because of different disease rates in different countries. In Japan, the incidence of breast cancer is much lower than in the West. But when Japanese families migrate to America, they take on the breast cancer rate of the United States within a couple of generations. "By 'environmental' we mean the whole of Western lifestyle. Something in the Western lifestyle has raised rates of breast cancer but we do not know what it is. . . ."[3]

Although the causes of this distressing disease are not understood, it is thought to involve many stages. "The first step in the development of the cancer is a process called *carcinogenesis,* the generation of cancer. Carcinogenesis begins with an initiating event which involves some sort of damage to the genetic code carried in the cell: the DNA," explains Philippa Darbre. "It also involves *promotion,* a process which enables those initiated or damaged cells to grow. During the next stage of *tumor growth,* there is a period of further growth of the cells into a lump that we can recognize. By the time there is a lump in the breast, if you look at the cells they have several characteristics. They are no longer identical copies of each other, some diversity has arisen: we call this *tumor progression.* Then, there is the process of *metastasis,* which is the spread of the tumor cells around the body to distant sites. Cancer cells can be transported either through the blood or the lymphatic system, which is involved with immunity. Usually, with breast cancer, it is the lymphatic system, so typically the breast cancer cells migrate to the lymph nodes under the arm and this is often the first place where metastasis is detected.

"There are studies that suggest that an initiated cell, with damaged DNA, can remain in an animal for the whole of its lifetime without doing anything and cause no problem. In other cases, the cancers may develop fast. . . . Since the body has a good immune system, there must be something special about cancer cells that they can actually get from one site to another and escape the normal defense mechanisms of the body. You would expect the body to detect a breast cell if it was not in the right place, but clearly these cancer cells are able to escape that mechanism."

In this extraordinarily complex multistep disease process, involving carcinogenesis (both initiation and promotion), tumor growth and progression, metastasis, and failure of immunity, the concern that estrogenic or other hormone-disrupting chemicals may play a role in one or more of these stages stems from several factors.

Firstly, explains Dr. Leffers, breast cells are *designed* to respond to estrogens. "During the menstrual cycle, the breast enlarges and then goes down again as estrogen levels change over the month. This is supposed to happen because breast cells are primed to respond to estrogens to prepare the body for reproduction, pregnancy, and so on." There are many estrogen receptors in breast tissue, and therefore, if estrogenic chemicals can act like "pass keys" and fit the estrogen receptor just like the woman's natural estrogen estradiol, then it is at least plausible they could exert effects.

Secondly, it is widely accepted that a woman's lifetime exposure to estrogen is an important risk factor for breast cancer. The longer the exposure, the greater the risk. The timings of critical landmarks in a woman's reproductive life can increase exposure and this correlates with breast cancer risk. Women who start their periods at a younger age or undergo menopause at a later age, and who have never had or breast-fed a child, are at an increased risk of the disease, and all of these factors are thought to increase lifetime exposure to estrogen. Women who have their ovaries removed at a young age and therefore have much lower exposure to estrogens have a much lower risk of breast cancer.[4]

Thirdly, as has been shown in Chapters 3 and 8, many of the man-made chemical estrogen mimics that are causing concern appear to target fatty tissue, such as breast tissue, and can be stored there, sometimes for years. In contrast, the natural estrogen estradiol usually only lasts a few minutes in the body and is then degraded into less potent forms and excreted. Such is the persistence and accumulation of certain environmental estrogens that they are typically measured in breast milk and breast milk fat in much higher levels than in blood. Some of the chlorinated environmental chemicals (which are the most stable because of the chlorine which makes them resistant to degradation), such as DDT and certain PCBs, are concentrated in fatty tissue, such as breast tissue, and can reach levels *200 to 300 times higher* in fatty tissue than in the blood, and much much higher than levels of circulating estradiol.[5] Little is known about what happens to these chemicals during storage, whether they are totally locked up in fat or whether they are available to affect neighboring cells. As has been shown earlier, during reproduction and lactation, and perhaps dieting (although this does not appear to have been tested), whenever fat supplies are mobilized, these chemicals can reenter the bloodstream.

Fourthly, it can be shown in the laboratory that some breast cancer cell lines will only grow in the presence of estrogens. Indeed, several of the tests investigating the estrogenic potency of a chemical are carried out using these breast cancer cell lines, such as Soto and Sonnenschein's work with MCF 7 cells, a strain that grows better in the presence of estrogens. This relationship between breast cancer cell growth and estrogens is also borne out by clinical practice. Dr. Philippa Darbre explains: "One of the things that has stood the test of time is that removal of estrogen can deprive the breast cancer cells of the factors they need for growth: if you remove the estrogens, some cancers will go into remission. . . . This is usually done with the drug Tamoxifen, which antagonizes the action of estrogen in the body, depriving the breast cancer cells of the estrogen they need to grow. In the clinic about 30 percent of breast cancers respond to such endocrine or hormone therapy; about 70 percent don't respond. For the 30 percent that do respond, this is a much nicer therapy than the standard chemotherapy or radiation because the side-effects are much less."

Finally, over the same timescale that hormone-disrupting chemicals have become widespread environmental contaminants there has been a startling rise in breast cancer. Fifty years ago *one in twenty-two* women would get breast cancer at some point during their lifetime.[6] "Breast cancer risk in the United States has been rising since 1940. There has been a steeper increase since the later 1970s and 1980s so that the risk in some parts of the country is now as high as *one in eight,*" explains Dr. Mary Wolff at the Mount Sinai

Medical Center in New York. "It's not clear why there's been a much higher breast cancer risk in the last couple of decades, some of it is due to more frequent screening of women by mammography, but there is a great deal of this risk that is yet unexplained."[7] In Britain the increase in incidence is almost as great, latest figures suggesting one in twelve women will be affected. In America, such is the concern about this dramatic increase that it has become a major public policy issue. "Hold the line, at one in nine," was the cry at one recent rally in Boston, where women were demanding more research for this devastating disease. "Now you can no longer afford to close your eyes," was the caption on a recent front cover of the *New York Times* magazine, featuring a beautiful young woman with her breast surgically removed, showing a huge scar across her chest.

There is, however, recent evidence suggesting that breast cancer mortality rates may be leveling off or beginning to decline in some Western countries. Dr. Valerie Beral of the ICRF Cancer Epidemiology Unit gathered data from the World Health Organization on breast cancer deaths in twenty countries. Although in Belgium, Hungary, Poland and Spain, rates are continuing to rise, in a number of countries, including Britain, there is evidence of a decline in mortality. She argues that better treatment and earlier diagnosis might be responsible for much of the fall in breast cancer mortality.[8] But despite the very recent decline in mortality, latest figures on incidence from the Cancer Research Campaign show an increase, although this too could be due to improved screening. "It's too early to tell from the figures whether actual incidence is still increasing or not," says Dr. Beral.

In summary, a number of provocative clues have led some scientists to question more closely whether estrogenic chemicals could play some sort of role in this very complex disease: firstly, breast cells are primed to respond to estrogens and have many estrogen receptors; secondly, lifetime estrogen exposure is a known risk factor for breast cancer; thirdly, estrogenic chemicals are stored in the body sometimes for years, targeting fatty tissue, such as breast tissue; fourthly, certain types of breast cancer will only grow in the presence of estrogens and can be controlled by antagonizing estrogen action in the body; finally, there has been a very steep rise in the incidence of breast cancer in the last fifty years, a timescale which just happens to coincide with the greatly increased production of these suspect chemicals.

The combination of all these factors does not *prove,* however, that chemicals play a role in the disease. Sadly, there is very little research in this important area. The few studies that have been done are controversial, but provide fascinating insights nonetheless.

Mount Sinai School of Medicine, New York

One of the first scientists to delve more deeply into a possible link between estrogenic chemicals and breast cancer incidence was Professor Mary Wolff at the Mount Sinai School of Medicine, in New York. In 1987, an intriguing study caught her eye, which suggested that DDE could be *hormonally active* in the breast.

Dr. Walter Rogan and his team from the National Institute of Environmental Health Sciences in North Carolina had undertaken a study of nursing mothers in which they found a curious relationship between levels of certain chemicals and duration of lactation. "Children of mothers with higher levels of DDE were breast-fed for markedly shorter times," they wrote. "We speculate that DDE may be interfering with the mother's ability to lactate. . . ." They found a decline in the duration of lactation of about one week for each additional ppm of chemical.[9] Lactation is under the control of a number of hormones, and it is unclear what the mechanism might be for DDE to interfere with this process. However, further studies have confirmed that DDE appears to shorten the duration of lactation.[10]

Perusing this data in the late 1980s, Mary Wolff began to question if DDE might in some way affect the duration of lactation in the breast, could it possibly have other hormonal effects in the breast as well? Knowing that DDE is the major residue of DDT and is found in humans at higher levels, and more frequently, than any other persistent chemical, she began to investigate further.

At the suggestion of her colleague Frank Falck, they designed a preliminary study to find out if women with breast cancer had higher levels of organochlorine compounds like DDT in the breast. They studied fat tissue from forty patients who had had a breast biopsy. Half of these patients had breast cancer, and half had benign disease. The results were striking. The twenty breast cancer patients had *50 percent higher* levels of DDE, DDT and the more chlorinated PCBs compared with the twenty controls of similar ages. In fact there was approximately 1 percent increased risk in breast cancer for every 10 ppb increase of DDE and PCBs in fatty tissue. However, conscious that this was just a pilot study, in the early 1990s she and her team undertook a larger study.[11]

She found out that at New York University Medical Center blood samples had been taken from 14,000 women for another purpose, as part of a general health study. By chance, of these women, fifty-eight had gone on to develop breast cancer within six months of taking part in the study. She wondered whether these fifty-eight blood samples might provide some further insights. To check this out, the fifty-eight women with cancer were matched for age and certain reproductive factors to controls, i.e. women who had not developed breast cancer. All the blood samples were sent to laboratories at Mount Sinai, where they were analyzed for levels of DDE and certain PCBs.[12]

"It took us about a year to analyze these blood samples," recalls Mary Wolff, "and of course we tested the samples without any knowledge as to whether a woman had breast cancer or not. Then we sent the results off to our collaborators in New York University who did the statistical analysis. They called me in the spring of 1992 to say that there was a very strong effect with DDT and breast cancer risk. In the final analysis it looked as though the women in the *highest 10 percent of exposures had about a fourfold* increased risk of breast cancer, which was quite surprising to us."

If those women with the highest levels of DDT in their blood had four times the risk of contracting breast cancer than women with lower levels, could estrogenic chemicals like DDT play a role in the disease? "There is something in those observations that's new and different that we need to understand in terms of breast cancer risk," explained Mary Wolff in an interview for *Horizon*. "There are multiple routes that may lead to breast cancer and it is certainly possible that DDT could contribute to one of these routes. . . . It's been suggested that the levels of synthetic estrogens are too low to make a difference, compared to naturally occurring estrogens that women already have. But in fact, the levels of DDE that we see today in women's blood *are 10 to 100 times higher than the naturally circulating estrogens* in our bodies. It's important to realize that these chemicals are not metabolized very easily. They stick around in the body for a long time, and therefore they can have a repeated effect."

In this study, Wolff and her colleagues found there was approximately 9 percent increased risk for every 1 ppb of DDE in the blood. PCB levels were also higher among some cases although the differences were not statistically significant. Her study immediately became headline news. "I believe the study received so much attention because the study design was very strong and these blood samples were gathered before women were diagnosed with breast cancer and they were very carefully matched for reproductive factors that we know are the other risks. Of course, we were able to measure exposure directly in the body, which is another very critical component in these kinds of epidemiology studies."

However, not all studies have found a link between breast cancer and levels of exposure to persistent contaminants such as DDE. During 1993, Nancy Krieger from the Kaiser Foundation Research Institute in California and her team collaborated with Mary Wolff on an even larger study to investigate the issue.[13] During the late 1960s, over 57,000 Californian women had taken part in a general health study in which blood samples from the women were frozen and stored to enable later follow-up. Krieger and her colleagues went back to trace these women and find out which ones had gone on to develop breast cancer. As before, the breast cancer patients were age-matched with controls who did not develop the disease. In all, 300 women were studied from three racial groups: one-third of the women in the study were white; one-third, black; and one-third, Asian. But, this time, there was no significant difference in levels of DDE or PCB in women who had developed breast cancer compared with controls. "The data do *not* support the hypothesis that exposure to DDE and PCBs increases the risk of breast cancer," Krieger and her team concluded. However, they pointed out that, even though this study was larger than earlier studies, "this study has several noteworthy limitations . . . the total sample size is still relatively small. . . . It is also possible that analysis based on frozen serum stored for almost twenty-five years may be unreliable. . . . Of more concern is the lack of adequate data on age at first pregnancy and lack of data on lactation since this is the chief route by which women excrete organochlorines. . . ." They recommended that future investigations should take account of a number of confounding variables.[14]

Professor Stephen Safe, a toxicologist from Texas A&M University, and others have cited the Krieger study to suggest that the proposed link between PCBs and DDE is questionable. "As of now, with the analysis of all the studies, including the big Krieger study, there is no evidence that organochlorine levels are higher in breast cancer patients versus controls. So that is in a sense a settled issue," he says. "This doesn't mean that the original data is wrong, it is just that for that small group they were higher. Could this be a dietary factor, because organochlorines are found in a fatty diet? I don't know. I think there is something there, but I don't think it is organochlorine related. . . . It is possible that these persistent chemicals are tracking different kinds of fat and that may be a problem. But that is just a hypothesis. It doesn't appear at the moment that there is a relationship between these compounds and breast cancer." In support of this view, he also cites occupational exposures, claiming that studies investigating women who have high exposures to PCBs and DDT in the workplace show no evidence of an increase in cancer risk.[15]

But, for others, it is far from a settled issue. "This study had 50 blacks, 50 whites and 50 Asians with breast cancer and it compared them to women who did not have the disease," explains Devra Lee Davis, former deputy health policy adviser to the American government. "When you added the Asians to the analysis the results were insignificant. But if you took the Asians out of the study something remarkable happened. The risk for breast cancer was two to three times greater for black and white women. That is to say, that the more pesticides in your blood, the greater your risk of breast cancer if you were black or white."[16] Professor vom Saal from the University of Missouri is also concerned about this: "The Krieger study included Japanese women and didn't ask them how long they had been in America. It takes essentially several generations for Japanese families to move from exceedingly low breast cancer risk to the level of breast cancer risk seen in the United States. If you take the Japanese data out of the study, it replicates the Wolff data."[17]

Others point out that Wolff's preliminary studies were not the only reports to show a relationship between breast cancer and certain hormonally active chemicals. In 1976, Dr. Wasserman and his team from the Hebrew University Hadassah Medical School, Jerusalem, Israel, working in conjunction with scientists from the World Health Organization, studied levels of DDT and PCBs in breast tissue. He compared breast cancer patients who were having breast tissue removed for medical reasons with tissue from women who had died accidentally. He also compared the concentrations of these compounds in the malignant breast tissue with those in adjacent tissues. The results were startling. "*The largest concentration of total DDT in cancer patients is found in malignant tissue,*" he wrote, "when compared with the adjacent mammary and adipose tissue. The same may be said about the concentration of PCBs, that is, the concentration of PCBs in extracted lipids of malignant breast tissue is *considerably higher* than in the normal breast and adipose tissue." This admittedly very small-scale study of just nine breast cancer patients would appear to suggest that comparing *overall* blood or fat levels of such persistent chemicals in relation to breast cancer risk may be too simplistic, since there may be individual differences in metabolism of these substances which may lead to higher concentrations in particular sites in the body.[18]

Interestingly, another study by Pascal Pujol from Montpelier in France, working together with a team of cancer specialists from the University of Texas, reported rising levels of *estrogen-responsive breast cancer* over the last two decades. In this large-scale study they analyzed the estrogen receptor content of over 11,000 tumors collected between 1972 and 1993. These were from patients diagnosed in hospitals across the United States. Specimens were sent to the University of Texas Health Sciences Center for estrogen receptor analysis. "This study found a steady increase in the measured levels of estrogen receptor in breast tumors in the last twenty years. This finding is consistent with two previous reports. . . . This report substantiates that the trend is continuing." After considering the factors that may be involved, they conclude: "This increase therefore may reflect a *change in tumor biology and the hormonal events that influence breast cancer genesis and growth.*"[19] This raises the issue of early sensitization of breast cells, which is discussed below in relation to prostate disease.

Another study on estrogen-responsive breast cancers also found a link with chemical exposure. Dewailly and his team from Laval University in Quebec found a difference in DDE levels between cancer patients and controls, depending on whether the cancers had high levels of estrogen receptor or not. "Our results suggest that women with hormone-responsive breast cancer . . . have a higher DDE burden than women with benign breast diseases," they reported. "This study supports the hypothesis that exposure to estrogenic organochlorines may affect the incidence of estrogen-responsive breast cancer." Such findings suggest that further studies of breast cancer risk need to take account of the type of cancer which is reported and whether it is a hormonally responsive type or not.[20]

Just to add to the confounding variables, there is also the question of how chemicals might interact in breast tissue. A Finnish study, from the Department of Public Health in Helsinki, found elevated levels of another persistent chlorinated compound, β-hexachlorocyclohexane (HCH), a residue derived from the pesticide lindane, in breast cancer patients compared to controls.[21] Wolff has pointed out that in addition to DDT and PCBs there are a number of estrogenic compounds to which we are exposed but which cannot be so readily detected in blood and fat, either because they are readily metabolized, or for other reasons. Consequently, women may be exposed to other compounds which may add to the effects, but epidemiologists will not be able to measure them because they escape detection. Bearing in mind that breast tissue may be exposed to a cocktail of different chemicals, attempts to link increased risk of breast cancer to individual chemicals could be of limited value.

Stephen Safe and others remain skeptical. He says that not only are elevations of persistent contaminants such as DDE and PCBs not consistently observed in breast cancer patients, but in addition to this the proposed mechanism is questionable. As shown in Chapter 9, recent studies by Earl Gray and his team have shown that DDE is *not* estrogenic. DDE, they found, appears to exert its feminizing effect, not by fitting the estrogen receptor, but by blocking the androgen receptor, the male hormone receptor. So how, Safe asks, could DDE be promoting estrogen action in the breast?[22]

Wolff and others acknowledge that there are still a great many unknowns, but argue that nearly all the studies do suggest a relationship between DDT exposure and breast cancer and other persistent chlorinated compounds such as chlordane, PCBs, hexachlorocyclohexane, kepone and PBBs: compounds "sequestered in fatty tissue and resident in the body for a lifetime." Furthermore several mechanisms can be proposed to explain this association, including their estrogenic activity, their effects on certain enzymes in the body which could activate other chemical carcinogens, and other complex mechanisms.[23] "It's not really clear how these chemicals might work to increase breast cancer risk, but the most likely role is as a tumor promoter. . . ." she says. "It seems to me it would be surprising if there weren't some kind of a link between environmental exposures of this kind with hormonal activity and reproductive cancer, since we know what a major role hormones play in these diseases. . . . The most important thing is that if these chemicals play a role in reproductive cancers then there is a means of preventing such exposures and therefore of avoiding the amount of cancer that comes from such exposures."

Currently, there are now some twenty studies under way in different countries to try to examine the question further, including some Third World countries where there is currently considerable exposure to DDT, such as in Mexico.

University of Reading, England

Apart from the insights gained from epidemiology, examining the patterns of human disease, there are further clues from laboratory studies. Dr. Philippa Darbre is one of the few researchers in this field in Britain. After nine years studying how hormones regulate breast cancer cell growth for the Imperial Cancer Research Fund, she moved to the University of Reading to work with a team who had specialist knowledge of the toxicology of PCBs, such as Professor Raymond Dils.[24] "Dils and others were studying the accumulation of PCBs in fatty tissue such as the breast and pointed out that PCBs are accumulating in breast fat and breast milk. That made me start to wonder what they are doing in the breast. Things started to twig . . ." recalls Darbre.[25]

She searched through the literature and soon found Professor John McLachlan's papers showing that certain PCBs may be able to bind to the estrogen receptor and could even prompt the growth of uterine cells, a classic test of an estrogen.[26] Intrigued that here was a group of toxic compounds, known to accumulate in fatty tissue, some of which could have estrogenic action, she wanted to investigate further. "One of the problems with PCBs is that there are 209 different forms and you can't test them all. So it is very difficult to know where to start," she says. "In my studies I just hit lucky really because I picked a couple of PCBs that were being investigated here at Reading with high toxicity and began to check these out in relation to breast cancer cell growth."

She started with one particular tetrachlorobiphenyl (TCB), a compound with four chlorine atoms attached to two joined phenol rings, which is known to be highly toxic and one of the more common PCBs to accumulate in body tissues, not least because of its stability and resistance to degradation. A number of studies have shown that PCBs can interact with one of the orphan receptors, known as the Ah (aryl hydrocarbon) receptor, which is thought to affect thyroid hormone metabolism.[27] Darbre began her studies by checking out whether TCB could, as McLachlan's study suggested, have estrogenic action as well.

In the laboratory, she was able to show that this compound could indeed bind to the estrogen receptor, just like estradiol. Not only this, but she was able to show that TCB could mimic the gene expression of estradiol, that is, it acted on the genes like the natural hormone. "Then we asked the more complicated question of whether it could make the breast cancer cells grow in the laboratory," she says. "When breast cancer cells are dependent on estrogen for their growth, could we put in TCB instead of estrogen and make the cells grow?"

Breast cancer cells were cultured in the laboratory. Tiny amounts of TCB were carefully added to the wells. Within a few days they had a result. "The answer was, it could. TCB could make the breast cancer cells grow just like estradiol, and at frighteningly low concentrations, only two orders of magnitude higher than the levels at which estradiol is active."[28]

She tried to compare these concentrations to the amounts we may be accumulating in our bodies. Here the data was confused. *Total PCB* accumulates in much higher levels than the levels of TCB she found would make breast cancer cells grow: up to 10 milligrams per kilogram of body fat. Although TCB is known to be one of the top seventeen PCBs to be accumulated in body tissues, she could not find figures on the concentrations of this individual PCB in body fat. "What is difficult to answer is, what is the concentration of a particular PCB near to breast cells in the body?" she explains. "One of my concerns is that there may be differences in different body tissues. . . . We can't make sweeping conclusions. But it makes me realize there may be something in it. We've obviously got to do a lot more work to try to find out how important this is."

Then she took her studies one stage further. It is, after all, one thing to show an effect on breast cancer cells in culture in the laboratory, but quite another to show that the compounds behave in the same way in mammals. "We took this same compound, TCB, and asked, if it can act as an estrogen, can it cause breast cancers? There is one standard model for studying breast cancer which has been used for many years. Rats are fed a single dose of a compound which will initiate mammary tumors, called DMBA (dimethylbenzanthracene). If given DMBA rats will develop mammary tumors but they develop at a slow rate, taking several months. You can then feed them another compound as well, and ask will it enhance the rate? We did this with TCB."

In this study there were control groups, which had no TCB, a group with TCB and a relatively low-fat diet of 5 percent corn oil, and finally a third group with TCB and a high-fat diet of 20 percent corn oil. Philippa Darbre was distressed at the result. "We found we had to terminate the experiment after just ten and a half weeks. We had a large number of mammary tumors in the group of rats fed TCB in conjunction with the high-fat diet. Most of the rats in this group had breast cancer at ten weeks. . . . *It did frighten me because with the high-fat diet and TCB combined, we couldn't even keep the study going for the normal length of time. It was worrying how fast the breast tumors grew.*"[29]

For obvious reasons she has not repeated this study, but is extremely concerned that the combination of the high-fat diet and the estrogenic chemical produced such marked effects. "To my knowledge this is the first time such a study has been done with PCBs on breast cancer cell growth. PCBs are pretty toxic compounds and they are normally tested on the liver. . . . We need to know a great deal more about what these substances may do in breast tissue. . . . Our work is only beginning to touch it." Asked if PCBs might be the answer, she adds: "It is very unlikely it will turn out to be one thing on its own. It is much more likely to be a combination of factors. The reason I say that is because in Japan they have low levels of breast cancer, but quite high levels of PCBs, the same as in the West. What is different about Japan is that they have a low-fat diet. Now you have to remember many of these hormonally active chemicals such as PCBs are fat soluble. You begin to get the message that we've got something that is very complicated, with a lot of interacting factors. Even if PCBs were the answer, it would have to go with other things like fat in the diet."

Strang–Cornell Medical Center, New York

Apart from the clues from the epidemiological and animal studies, other research is in progress trying to understand in much more detail what happens to the natural female hormone estradiol in the body and whether this may be altered during breast cancer. Once again, recent research shows that estrogenic chemicals may play a role in this process too. A controversial theory which has been gaining increasing momentum in America has been proposed by Dr. Devra Lee Davis, Professor Leon Bradlow and colleagues at the Strang-Cornell Cancer Research Laboratories. They suggest that the events that lead to breast cancer may depend on how estradiol is metabolized, or broken down, in the body. Recently they have put forward evidence suggesting that certain hormonally active chemicals may interfere with this process and thereby exacerbate breast cancer.

In an interview for *Horizon,* Dr. Michael Osborne, formerly from the Royal Marsden in London, but now director of the Strang-Cornell Medical Center, explains how the ideas have developed. "Knowing that lifetime estrogen exposure is important in the development of breast cancer, our team explored every possible way in which estrogen may be involved: the concentrations of estrogens, the breakdown products of estrogens, known as estrogen metabolites, and all the hormones that come out of these breakdown products. It was a very lengthy and tedious process, but eventually we found one hormone, an estrogen breakdown product, that appeared to be abnormal and had some very interesting properties. We called it: 16α."[30]

This breakdown product of the female hormone estradiol, known as 16α-hydroxyoestrone, appeared to have some undesirable characteristics. "What is intriguing is that this particular hormonal breakdown product, the 16α form, appears to have the ability to bind irreversibly with the genetic material of the cell, the DNA," Dr. Osborne explains. "That gives it the potential to damage the genes, switch them on or change them in some irreversible way."[31] Using an elaborate marking technique, they measured the rate of production of this breakdown product, 16α, in patients with breast cancer and those without. "We were very surprised at what we found," recalls Dr. Osborne. "In women who had had breast cancer, in their nearby normal tissue there was a very large increase in the level of production of this breakdown product, the 16α form, compared with women who had not had breast cancer and who were having surgery for a harmless condition."[32] They also report an increased production of the 16α form in relatives of patients with breast cancer and in animals that are prone to develop breast cancer. Further tests revealed that several known risk factors for breast cancer are associated with increased levels of production of 16α, such as increased weight, a high-fat diet, and high alcohol consumption.

"Then we began to wonder whether environmental chemicals that have been linked to breast cancer could have any effect on the formation of this hormone," he explains. "When we looked, we were intrigued to find that DDT can elevate the 16α form and DDT could well be a potential risk factor for breast cancer. . . . So we went on to test many environmental chemicals, including the pesticides atrazine, kepone and DDE, and we found a similar effect for all of these. They changed the way estrogen was broken down into the 16α form, which we believe may promote the development of breast cancer." In their report of this study they found an increase in the formation of the 16α with the following environmental chemicals: DDT, atrazine, benzene hexachloride, kepone, certain PCBs, endosulfans I & II, linoleic and eicosapentenoic acids, and indole-3-carbinole. The DDT-related compounds had the greatest potency. "A number of lines of evidence suggest that 16α is a biological marker of risk for breast cancer," they wrote, "and may directly contribute to the initiation and progression of the disease."[33]

"We're very worried that a lot of the chemicals in the environment that we've just taken for granted may actually have an effect on breast cancer," warns Dr. Osborne. "Women *may* be developing breast cancer because of exposure to these chemicals. It's very urgent to explore this further and find out whether they really are involved and whether we can do something about it. This realization has only just recently come upon us and I think it has surprised us and indeed may explain some of what we don't know about breast cancer. It's an exciting finding."[34]

If this is correct, it highlights the complexity of the factors involved, suggesting that environmental chemicals may interact with the natural female hormone to alter the way this is broken down and metabolized in the body. Dr. Devra Lee Davis, visiting professor at the Strang-Cornell Cancer Research Laboratories, acknowledges that this view is not universally accepted, but argues there is enough evidence to warrant concern. "Although exposures can be very small, truly microscopic, the trouble is we are exposed throughout a lifetime and fat attracts these materials like a *natural hazardous waste site,* so that by the time women get to their fifties, when they show an increased risk for cancer . . . one of the things that may be increasing is their exposure to these kinds of harmful compounds," she says.[35] "Cancer is a complicated disease, resulting from many interacting factors that may differ from one person to the next. We realize that environmental estrogens cannot account for all breast cancer . . . but if reducing avoidable exposures made it possible to avert only 20 percent of breast cancers every year . . . at least 36,000 women (in America) would be spared this difficult disease. . . . Such prospects are too tantalizing to ignore."[36]

Early Sensitization: Prostate Disease

University of Missouri, Columbia

Apart from breast cancer, it has already been shown that several other reproductive cancers may be affected by hormone-mimicking chemicals. Firstly, Skakkebaek's studies showed how testicular cancer could have its origins in the womb,[37] and how estrogenic chemicals may play a role.[38] Tragic case studies of human exposure to diethylstilbestrol (DES), a potent estrogen which was widely given in the 1950s and 1960s to prevent miscarriage, have shown it is critical in the development of a rare vaginal cancer in exposed daughters. Animal studies with DES have suggested the same mechanism may be involved in prostate cancer.[39]

Recently, at the University of Missouri, Professor vom Saal has carried out further studies on prostate disease. Many of the early studies were carried out with large doses of the synthetic estrogen DES. Yet Professor vom Saal's own studies on mice in the womb had shown tiny differences in exposure to hormones, arising simply from whether a fetus was positioned next to a male or a female, could produce changes in behavior and reproduction in the adult. This made him concerned that such tiny increases in exposure to estrogens prenatally could *permanently sensitize* organs such as prostate glands in the womb, predisposing to prostate disease later in life.[40]

To study such sensitization further, in studies on mice, male fetuses were given unbelievably tiny increases of either the natural hormone estradiol or the synthetic estrogen DES. "We increased the dose of estradiol by one tenth of a trillionth of a gram per milliliter of blood," he explains. "This resulted in a 40 percent increase in the number of developing prostate glands, detectable on the first day of prostate development in the embryo, which is a substantial increase in prostate gland development. . . . What we then found was, in adulthood, these animals had on average a 40 percent enlargement of the prostate and a smaller urethra. They also had twice the number of receptors for male sex hormones (androgens) than males not given this prenatal increase in estrogens. *Estrogen had programmed the organs in the treated males to be hypersensitive to male sex hormones for the rest of that animal's life.* So estrogen exposure before birth produced a double effect: an enlarged prostate and a smaller urethra. And what puts men into a doctor's office?" he asks. "That's it! When they can't urinate, because the prostate is enlarged and pressing down on the urethra."[41]

Studies with the synthetic estrogen DES produced the same result. These studies suggest that increased exposure to estrogens in the womb may permanently alter the way the prostate develops, predisposing to prostate disease later in life. "We can't say that this is the basis for some men having a somewhat greater tendency than others to develop prostate enlargement," he says. "But anyway it is intriguing."

What is also intriguing is that the tiny increases in hormonal exposure before birth resulted in *doubling* the number of male hormone receptors in the organ in adulthood, creating a permanently increased sensitivity to hormonal exposure. If such a mechanism plays any part in breast cancer, it would certainly help to explain why the epidemiology is so confusing, since prenatal exposures to synthetic chemicals may be more important than current exposures. "The embryo has a hormone system that is in the process of developing, and the consequences of exposure to hormone-disrupting chemicals at one time in the maturation of the embryonic hormone system can be totally different than the consequences of exposure at another time," he warns. "The capacity for tissues to respond to hormonal signals changes during development, and the effect on a tissue of exposure to hormone-disrupting chemicals will change throughout development."[42]

Vom Saal highlights another factor in his study which would confuse any epidemiology. With any increases in estrogen exposure before birth, they had marked effects later; yet, strangely, as the dose went up, the effects on the prostate were less. The dose-response curve is like an inverted U: if you give too big a dose, you get nothing. "That's toxicology for you," he says. "Now everyone else is back in the lab, with much lower doses."[43] If this result is confirmed in other laboratories, it could provide an important model for prenatal exposures to hormonally active chemicals.

At present, however, none of the studies *prove* that environmental chemicals are behind the increase in such diseases and cancer. However, many scientists are worried at the accumulating evidence of a link between human health and exposure to hormonally active chemicals. "Some people require dead bodies before they believe there is an association between an exposure and effect," argues Devra Lee Davis. "I think as a matter of sound public health we should not insist on proof of dead bodies. What we have to do is pay better heed to experimental studies, to wildlife studies, and now to the growing human evidence that we have, all of which point in the same direction. All these studies suggest that there are widespread exposures to some environmental chemicals that disrupt hormones and that some of these could be behind the increases in breast cancer, testicular cancer and perhaps prostate cancer. I would argue that we can't afford to ignore this evidence."[44]

In summarizing what could be at stake and the choices that we have, Dr. Devra Lee Davis likens this situation to a bargain with Faust. "It's as though we have unwittingly struck the ultimate Faustian bargain. In return for all the benefits of our modern society, and all the amazing products of modern life, we have more breast cancer, more testicular cancer and reduced fertility. But I don't think any young woman who has lost her breasts, or any young man who wishes to be a father and cannot, would willingly endure this for the sake of modern society. It's a price that is too high for anyone to pay."

But in this Faustian bargain, is science able to provide an exact measure of the human price? In the balance on one side are the reproductive cancers and the decline in sperm counts, with all the uncertainties and unknowns associated with this data. In the balance on the other side appear to be the countless luxurious products of modern living to which we are so accustomed that we take them for granted. But is this the complete equation? Have we got the full measure of this Faustian bargain? There are those who argue the human price is higher even than this, and goes way beyond the controversy over sperm counts and cancer of the testes, breast and prostate. Some have argued that much more fundamental aspects of humanity are also being subtly and insidiously eroded.

The Price in Full

Taiwan, 1979

Two thousand people were caught up in a strange epidemic, suffering signs of chronic poisoning. Eventually a diagnosis was made. They were suffering from "oil disease." Rice oil was accidentally contaminated with high levels of extremely toxic substances, principally a complex mixture of PCBs and a close chemical relative, the furans.[45] During this time over a hundred children were exposed to these chemicals, either in the womb or through their mother's milk. These children were studied for a number of years by a scientific team from Taiwan's Department of Occupational and Environmental Health, and dramatic adverse effects were found.[46]

Firstly, there were physical symptoms: a greater number of stillbirths, retarded growth, respiratory difficulties and reproductive abnormalities in the exposed babies. In addition to these changes, the scientists observed other subtle effects as well. In a series of tests the children exposed to PCBs showed differences in mental ability: developmental delays and lower intelligence, typically scoring 5 points less on intelligence tests than unexposed children. The most marked adverse effects were in first and second children born to exposed mothers.[47]

These results mirror a similar poisoning incident eleven years earlier in Japan. Cooking oil had been contaminated by PCBs used in decolorization during the processing of the oil. Over a thousand adults were affected. Children who were exposed in the womb suffered a number of physical and developmental problems. Followup studies revealed growth impairment, "an average IQ of 70, sluggishness, clumsy and jerky movement, and apathy."[48]

This research on infants accidentally contaminated with high levels of PCBs prompted investigators to find out if lower-level exposure could also affect mental development and intelligence. The first such study was begun in the eighties on children whose mothers had consumed Lake Michigan fish, which were among the most contaminated in the United States at the time. Sandra and Joseph Jacobson, two psychologists from Wayne State University in Detroit, studied over 300 infants. In the "exposed" group, all the mothers had eaten moderate amounts of Michigan fish, two to three fish meals a month, in the six-year period before becoming pregnant, amounting to an average of over 11 kilograms of fish. The controls were infants whose mothers had not eaten any fish. Prenatal exposure was assessed by taking a blood sample from the umbilical cord. Greater amounts of PCBs were transferred during breast-feeding and this was assessed by measuring the contaminants directly in the milk.[49]

Even though the amounts of PCBs consumed were much lower than in the accidental exposures in the Far East, the Jacobsons observed significant differences between the two groups. These differences were apparent in the newborn babies: the higher the mother's consumption of fish from the lakes, the smaller the circumference of the head and lower the birth weight of the baby. "Our study is the first to assess the effects of PCB exposure at ordinary dietary levels on the human infant," they wrote. "Exposed infants were up to 190 grams lighter than controls, a magnitude comparable to that reported by the Surgeon General for smoking during pregnancy."[50]

But what caused greater concern were the signs of impaired mental and learning ability as the children developed. As newborns, the exposed group showed abnormally weak reflexes and less well coordinated movements. At seven months, the exposed infants appeared to show poorer "visual recognition" memory. In this test, the infant is shown two identical photos for 20 seconds. This is thought to permit some kind of encoding of the image into short-term memory. Then one of the photos is replaced with an image of a new face. Normally, most babies will spend more time looking at the new photo. From this it is inferred that the baby has formed some kind of "memory" of the initial photo, or stimulus, which the baby can retrieve, so they can recognize that the new photo is indeed a new stimulus. But those children who had greatest exposure to PCBs did not respond in this way. Essentially they showed no preference between the new and the old face or stimulus, which suggests they did not recognize that this was indeed a new face. Subtle alterations such as these persisted throughout childhood: the infants exposed to higher levels of PCBs performed less well in a variety of developmental tests.[51]

These studies have been criticized, partly because of the difficulties of assessing something as complex as intelligence and development, partly too because there is always a risk of spurious correlation, that is, the observed relationship may be due to a third unmeasurable influence rather than exposure to PCBs. However, despite the difficulties of research in this area, others have reported similar findings. Walter Rogan and his team in North Carolina, in a study on over 800 infants, also found those infants exposed to higher levels of PCBs in the breast milk had poorer reflexes as newborns, and performed less well in a variety of tests on motor coordination.[52] Recent studies of children exposed to contaminants through mothers eating Lake Ontario fish have also found changes to aspects of cognitive development.[53]

Apart from alterations to learning and intelligence, other effects are being studied. As outlined in Chapter 9, Earl Gray and his team have shown how endocrine-disrupting chemicals can target not just reproduction by alterations to the sex hormones, but can also interact with thyroid, adrenal, and other hormone systems. Research is under way to investigate this further. Abraham Brouwer and his team from the department of toxicology at the Agricultural University, Wageningen, in Holland, have been investigating immune changes. Infants from the industrialized Rotterdam area were compared to infants from the semi-urban Groningen area in northern Holland. They found significant differences in thyroid hormone levels in the group of babies with higher exposures to PCBs and related compounds in human milk.[54] Such findings are also borne out by animal studies; for instance, research on seals has shown that those fed a diet of more contaminated fish had impaired immune function and were less well able to fight off viral infection compared with those seals eating relatively uncontaminated fish.[55] Similar immune changes have been reported in dolphins contaminated with higher levels of PCBs.[56] Studies on behavioral alterations such as increases in aggression are also in progress. Ironically, animal studies suggest that rodents exposed to excess estrogens in the womb show more aggressive behaviors later on.[57]

All these findings have prompted some to argue that the effects, although subtle and hard to prove, may be much more pervasive than previously thought. A recent conference in Erice, Sicily, brought together eighteen specialists in neurological and behavioral effects of hormone disruption, including Dr. Earl Gray, Dr. Abraham Brouwer and Professor Frederick vom Saal. At the meeting, they arrived at the following consensus:

Hormone-disrupting chemicals can undermine neurological and behavioral development and the subsequent potential of individuals exposed in the womb. . . . The developing brain exhibits specific and often narrow windows during which exposure to hormone disrupters can produce permanent changes in its structure and function. . . . A variety of chemical challenges in humans and animals early in life can lead to profound and irreversible abnormalities in brain development at exposure levels that do not produce permanent effects in an adult. . . . This may be expressed as reduced intellectual capacity and social adaptability, as impaired responsiveness to environmental demands or in a variety of other functional guises. Widespread loss of this nature can change the character of human societies. . . . [58]

"What we fear most immediately is not extinction, but the insidious erosion of the human species," argue Colborn, Dumanoski and Myers in *Our Stolen Future*. "We worry about an invisible loss of human potential. We worry about the power of hormone-disrupting chemicals to undermine and alter the characteristics that make us uniquely human. Our behavior, intelligence and capacity for social organization. . . . *What could be at stake is not simply a matter of change to some individual destinies, or impacts on the most sensitive of us, but widespread erosion of human potential.*"[59]

CHAPTER TWELVE

The Response

In America in 1993, many of the scientists at the forefront of the research were invited to testify before Congress. Panels were convened. The press were invited. Ana Soto, John McLachlan, Louis Guillette, Theo Colborn and others flew to Washington, D.C., and assembled in the State buildings on Capitol Hill. Professor John McLachlan, who had been one of the first to warn of possible adverse effects from environmental estrogens more than twenty years previously, found himself with fifteen minutes to summarize the significance of a lifetime's work. "I gave my testimony in terms of what environmental estrogens are, how they work, what a receptor is, and what we knew from experimental evidence in the laboratory and humans," recalls John McLachlan. "The Congressional staff and their attendants were not, I thought, overly interested. In fact hardly anyone was there! I don't think the room was completely empty. At all times someone was listening, maybe one or two congressmen and their staff, but they were usually doing something else at the same time. I do remember at some point I diverged from my written testimony to try to explain what an estrogen was. I was using my hands to explain a lock and key and so on. Suddenly I realized there wasn't anybody paying any attention to what I was saying. All the visual aids were wasted."[1]

He stopped, and the recorder clicked to a stop. In the distance, the sound of traffic could be heard, circulating around the State buildings and Capitol Hill. Inside was the hush of whispered conversations dealing with the urgent political business of the day. There were no questions—just silence—as he stepped down from the microphone. "I don't remember feeling that it accomplished anything," recalls McLachlan. "Not much seemed to happen."

But although the response seemed muted at the time, their voices *were* being heard. In fact, it's difficult to envisage a set of data coming from scarcely a dozen or so independent scientists which was to have such worldwide impact. However, even though the response was eventually considerable, not all of it was necessarily going in the direction the scientists would have wished.

The Response: Industry

The official response of the chemical industries and other manufacturers involved, is one of care and cooperation. A statement released from the European Chemical Industry Council (CEFIC) said: "The chemical industry shares public concerns and recognizes the importance of addressing any issue related to the safety of its products. Through the European Chemical Industry Council, chemical manufacturers are working with the scientific community and regulatory authorities in Europe and America. Our goal is to determine whether there is a deterioration in reproductive health that can be attributed to the influence of man-made chemicals on the hormonal activities of humans and wildlife. . . . An exhaustive review of available research is presently being made. . . ."[2]

A large research program has been set up, funded by industry. In America, the Chemical Manufacturers Association, together with other trade associations, is financing numerous studies, including an investigation of the link between DDT and breast cancer. A panel of experts has been convened to evaluate trends in semen quality and a study has been set up "which will correct for the flaws the expert panel members identified in existing studies on semen quality."[3] Research is under way to determine if is there is an association between PCBs and DDT and endometriosis. Additional studies are in progress on wildlife, dioxins and toxicology, screening programs, and many other related topics.[4]

"There are people in industry, particularly toxicologists and other research scientists, who take the issue very seriously indeed," says Professor Sumpter. "Some of the people I have met through the European Centre for Ecotoxicology and Toxicology of Chemicals in Europe have been the most balanced and well-informed people on the issue that I know of. Just as in academia, there are people in industry who are for and against the issue."

The Chemical Industries Association in London has a membership of 230 companies, many of which are international, and annual sales of £35 billion.[5] Summing up the considerable research effort on the part of industry, their spokesperson told BBC *Horizon*: "Chemical industries across the world, together with leading academics, are pursuing research in this area very, very *vigorously*. As a scientist, I have not seen such a vigorous attempt to get a clear answer in partnership between academia, government authorities and industry before. This is a very big exercise that is going on and it is being pursued by the world experts. . . . There is not a government in the developed world that doesn't have a research program on this."[6]

But just how "vigorous" have all the interested parties been? For those dealing with industry, there is also evidence of a different response behind the scenes.

"Industry doesn't bury its head in the sand," explains Dr. Richard Sharpe. "They may try that initially, thinking this will go away because there is nothing to it. Once it doesn't go away, they become quite proactive. I think what that proaction involves is actually two things. One is to tackle the issue the way I think it should be addressed, which is to fund some research to find out about the problem. But the other is to put into place delaying tactics, because you're talking about big money. The longer they can delay the solution to the problem, not only does it mean they have saved themselves a lot of money, but it also gives them time to get alternatives in place so they don't lose out in the market."[7]

An interview with a senior member of the chemical industry in America, who does not wish to be identified, confirmed the significance of delays. He expressed concern and dismay at some of the evidence which has been gathered, and told us: "Most companies spend a tremendous amount of money on safety and the environment. Nonetheless they do not wish to disturb the smooth running of business. Behind the scenes there is a race going on and very extensive work. But this is all being done very quietly. The real change you don't hear about. The process of change is as follows: industries will nonchalantly agree with regulators that we are not to sell this product or that product at some point in the future and then allow enough time to find alternatives which they have tested. The pace of change is not determined by governments or politicians. It is determined by industry. This is the way the system works. Everything is happening very quietly because if the implications of this are that certain chemicals may play a role in the development of cancer, then because of the nature of the law in America, anyone who has cancer may put in a claim and this might open the door to frivolous lawsuits."[8]

So how might such delaying tactics work? "Industry is actually honest," Sharpe believes, "but dealing with them is like dealing with politicians. They will answer a question if you ask it, but they will never actually volunteer anything. So if they know you are trying to find out something, they will answer each of your questions, but they won't tell you what the question is you should ask them, to get the answer you are after. For instance, they might tell you about each of the chemicals you ask about, quite accurately. However, they don't tell you that in fact the most important route by which you are exposed to a given chemical is when they are conjugated or metabolized into something else. This is the difficulty, getting around all this. . . ."

"To try to give you some idea of how difficult this is for scientists and the scale of the problem," explains Professor Sumpter, "if in our group we identify a single new chemical that is estrogenic, then we need to know several things. Firstly, is this a single chemical, or a member of a chemical family? In some instances, the phthalates for example, are a family of about forty or fifty different chemicals, so we need to know what those chemicals are and whether they also are estrogenic. Secondly, we need to know what those forty or fifty chemicals actually degrade to, because we have to test the degradation products as well as the chemicals themselves. In addition to this, we need to know what products these chemicals are used in, otherwise we cannot derive information about possible exposure levels. So from a scientific point of view, the whole problem is almost insurmountable unless you have help from industry telling you about the chemicals, what related chemicals are in use, how they degrade, how much they are used, and what products they are in. . . . As scientists, if we antagonize them, then all that expertise is lost to us. We simply can't address the issue as academic scientists without industry. We don't know enough about the chemicals. We don't know how they are used, how they are made, how pure they are, the levels of exposure, and so on."[9]

But apart from sometimes being less than forthcoming with relevant information which only industry can possibly know, manufacturers can also take recourse to "trade secrets" and simply refuse to give information even when asked. This was Professor Soto's experience, when on the trail of nonylphenol, as outlined in Chapter 7. Some scientists working in the field have been genuinely concerned about coming up with results which will antagonize industry, since this, in some cases, can lead to difficulties with sources of funding. In addition to all these pressures, there is evidence that industry can delay advances in understanding in other ways as well.

"I was surprised to read, in an American industry magazine, information about my work," says John Sumpter. "The article said. 'Professor Sumpter's work has been replicated in the U.S. and the researchers *cannot* confirm his findings!' " Professor Sumpter knew of no such group. "Later, two people from industry whom I have been dealing with faxed me and said, 'Look we are very sorry about this. We don't know how it happened. It wasn't us, basically. . . .' Well, in fact there is no such work in the U.S. and all the research that has been published has supported my findings. . . . This kind of thing is not remotely constructive."

Some scientists, such as Dr. Theo Colborn, are concerned that the public relations exercise on behalf of industry is considerable and has a significant influence on how these issues are communicated to the public. Her work, both in organizing the Wingspread Conference and her recent book *Our Stolen Future*, written with Dumanoski and Myers, outlining possible health risks of hormone disruption, has been studied by industry committees. The Endocrine Issues Task Group, in a strategy paper for the Chemical

Manufacturers Association, assessed the likely impact of *Our Stolen Future* prior to publication. "The book is anticipated to contain an inventory of suspected endocrine-related effects of chemicals in animal species and call for additional research to identify causal agents. Previously published articles by Dr. Colborn have contained a list of approximately forty-eight products or environmental contaminants which she claims cause estrogenic, anti-estrogenic, androgenic, anti-androgenic, or other endocrine-mediated effects. Publicity associated with the book is expected to result in renewed media attention to the endocrine issue."[10]

Within industry there is anxiety that such adverse publicity could create widespread fear of chemicals, or "chemophobia." Several manufacturers' associations have formed an alliance to work together on the issue of hormone disruption. For example, a summary of agreements of the "Inter-Association meeting on Endocrine Issues" in Washington in September 1995 shows that the meeting was attended by representatives of the following associations: the Chemical Manufacturers Association, the Chlorine Chemistry Council, the American Plastics Council, the Society of the Plastics Industry and the American Crop Protection Association.

The agreements include the following:

- The existing Endocrine Issues Coalition (EIC) will serve as the focal point for developing a *coordinated cooperative strategy* for our associations on endocrine issues regarding:
  communications, including a process for coordinating responses to media inquiries;
  advocacy at the state, federal and international levels;
  scientific research, including when appropriate, joint funding . . .
- A communications task group will be formed. . . . The initial task of this group will be to develop a common standby statement, with umbrella messages to be used by all Endocrine Issues Coalition (EIC) members. In addition, more detailed statements addressing issues unique to each association's members will be developed and shared. . . . A process for referring media calls to one another will be developed, including contact points within each association . . .

- A separate advocacy task group will not be formed at this time. Advocacy coordination will be initially handled within the communications subcommittee.

At the meeting it was also agreed to set up additional committees, including the Endocrine Issues Coalition steering committee and steering committee coordinating group. Plans were set in place to approach other relevant trade associations to join the coalition.[11]

More recently, a memo from the Endocrine Issues Task Group to the Environment Health, Safety and Operations Committee of the Chemical Manufacturers Association has reported on industry's legislative concerns. "In the last Congress, endocrine issues played a prominent role in the push for restrictions on chlorinated compounds, and legislative proposals were offered by Senator D'Amato and others to require broad-based testing of chemicals for their estrogenic potential. Despite the overall focus of this Congress on regulatory reform, endocrine issues have remerged as a legislative priority."

Manufacturers fear that the already substantial data on chemicals which persist in the environment and human tissues will be put together with the new data emerging on hormone disruption to make the need for much tighter legislative controls even more urgent. "A key trend in regulatory thinking inside and outside the U.S. is the convergence of endocrine issues with the growing pressure to control persistent toxic bioaccumulators (PTBs)," reports the Endocrine Issues Task Force. "Internationally, numerous governmental organizations . . . are currently debating PTB policy. The hypothesis that PTBs could accumulate to levels that could cause reproductive and developmental abnormalities via a hormone-related mechanism is lending urgency to the perceived need of governmental bodies to limit release of possible PTBs (such as PCBs and dioxins) into the environment."[12]

To tackle this and other issues, among the action plans it was agreed that forming "global partnerships should be a high priority. . . . Global partnering will be of value both to leverage resources in conducting research . . . and to take full advantage of an aligned, harmonized approach to risk assessment and data interpretation by the global chemical industry."[13] American trade associations are now working with their European counterparts on these issues.[14]

It is difficult to capture fully the context in which this particular cutting-edge science is being produced, where a few independent scientists are coming up with data which inadvertently undermines one of the most powerful economic forces in society today. Most of the scientists involved are working with limited resources and backup, largely off their own hunches and interests. By contrast the vast enterprise of business can summon highly paid lawyers, public relations experts, and millions for research and development. Dr. Colborn and her colleagues point out that the chemical industry is such a powerful force in the global economy, sales of synthetic chemicals and products derived from them constitute well in excess of a third of the world's gross national product.[15]

The Response: Governments

While press releases, position papers, steering committees, coordinating committees, and countless others continued to claim that all was well, faced with mounting public concern, many Western governments commissioned their own inquiries into the issue. Of these, the most hard-hitting report came from Denmark.

Funded by the Danish Environmental Protection Agency, this independent committee concluded: "It is now evident that several aspects of male reproductive health have changed *dramatically* for the worse over the past 30 to 50 years. The most fundamental change has been the *striking* decline in sperm counts in the ejaculate of normal men. . . . Many otherwise normal men now have sperm counts so low that their fertility is likely to be impaired."[16] The report also found that, although the causes underlying these apparent changes are currently not clear, "both clinical and laboratory research suggests that all the described changes in male reproductive health appear interrelated and may have a common origin in fetal life or childhood. This means that the increase in some of the disorders seen today originated 20–40 years ago, and the prevalence of such defects in male babies born today will not become manifest for another 20 to 40 years or more." The role that estrogens might play in many, if not all, the reproductive disorders was cited, and it was pointed out that *"the large number of chemicals in numerous environmental categories suggests adequate availability."*[17]

The panel of experts identified many research priorities. In particular, it was found more work was urgently needed to assess: to what chemicals man is exposed; by which routes and to what degree; whether the chemicals get absorbed; the concentrations of these chemicals in man and how they are distributed in the body; whether they are mobilized in certain states such as pregnancy and, if so, how? The report ended rather more pessimistically with: *"Presently, we do not have adequate answers to any of these questions mainly because we do not know what chemicals are of greatest concern."*[18]

In London, an independent review was also commissioned by the government, which was rather more British about the whole thing. Scientists at the Institute for Environment and Health in Leicester prepared the report, which pointed out the lack of proof. They found: "There is compelling evidence that testicular and female breast cancer rates have been increasing during the last four decades in Westernized countries."[19] However, they pointed out, "as yet a *causal* relationship between exposure to environmental estrogens and adverse effects on reproductive health has not been established," although such an association must now be regarded as "plausible." They did acknowledge that "proof of a cause and effect relationship between exposure to estrogens in the environment and adverse effects on human reproductive effects is likely to remain *elusive."* They too prepared detailed research recommendations.[20]

Apart from Britain and Denmark, the German and Dutch governments have both recently initiated reviews and a European-wide research effort has now been started through the European Centre for Ecotoxicology and Toxicology of Chemicals. This is investigating screening techniques for hormonally active chemicals and regional differences in adverse health effects and other studies. In America, several government agencies are involved. The Environmental Protection Agency has launched extensive research on hormonal effects and identified the issue as a top research priority in its strategic plan. Studies under way include research on breast cancer, screening methods, further testing on hormonally active chemicals and many others. The National Academy of Sciences has convened a panel of experts to assess the data on hormonal effects. Several of the American contributors to this book are taking part, including Professor Ana Soto and Professor Stephen Safe.

However, in the absence of definitive proof, as yet all governments have stopped short of implementing legislation to protect the public from hormonally active chemicals. One of the problems for policy makers is reflected in the British government report: while proof remains elusive, it is very difficult to allocate significant resources to the issue.

Some journalists and environmentalists in Britain have argued that the government is not following its own guidelines.[21] Given the difficulties of establishing scientific proof on environmental issues, governments have endorsed the "precautionary principle" when weighing up scientific uncertainties, in which action may be taken before causality is established beyond doubt. This was outlined in the government's 1990 publication of Britain's Environmental Strategy which stated: "Where there are significant risks of damage to the Environment, the Government will be prepared to take precautionary action to limit the use of potentially dangerous materials or the spread of potentially dangerous pollutants, even where scientific knowledge is not conclusive, if the balance of likely costs and benefit justifies it."

Despite the laudable intentions, the precautionary principle has not been applied to protect the public from the potentially adverse effects of hormone-mimicking chemicals. Government departments have not specified what evidence they require before they will take action. No comprehensive strategy has been outlined for encouraging businesses to switch to substitute materials or for ensuring that the chemicals that are in commerce are systematically checked.

"The problem is, can we afford the precautionary principle?" asks Dr. Richard Sharpe. "The people who planned it probably never thought through what it might actually mean. It's okay to be doing it prospectively for a new chemical. But it is completely different when you've got a chemical out there and it's part of the fabric of society and then you say, 'Let's apply it to that.' . . . Most people's reaction would be, why don't we just simply ban all these chemicals if they are under suspicion and that is the end of our problem. I wish it were that simple. But the chemicals that we are talking about are used in so many of the products that are part of our modern everyday life that to get rid of them would actually introduce a revolution, the like of which we have not seen before in our way of living."

"Suppose, for example, we were to get rid of phthalates," says Professor John Sumpter. "You'd lose half the items in your house. The furniture, the PVC plastic, items like the washing machine, fridge, freezer, and so on, these all have lots of plastics. Then most of the cosmetics would go, food containers and wrappings. Do we really want this? This is not to say we shouldn't be mounting pressure on manufacturers to change and improve and taking action to phase chemicals out if we know we have a genuinely safe alternative. But it needs thinking through."

So despite the flurry of government reports, and considerable research efforts, no steps have been taken. Once again, it would appear the public is caught in a catch-22 situation. Governments are unlikely to introduce protective measures in the absence of proof, yet obtaining definitive proof, as stated in the British report, is going to be elusive. "I think it would be difficult, if not impossible, to get absolute proof," says Sumpter. "Some philosophers of science argue that you never, ever prove a hypothesis. What you do is accumulate evidence that disproves it. At the moment what we are doing is accumulating the evidence, and most of the evidence supports the hypothesis that exposure to estrogenic chemicals affects the testes and sperm production, so we're getting closer and closer to being able to say this hypothesis is probably true. But there will always be an argument as to whether you absolutely definitively prove it. So it is a 'weight of evidence' argument, and it seems to me that all the evidence is slowly but steadily going in one direction."[22]

The question is, what constitutes a "weight of evidence" and for whom?

The Response: The Scientists

While for some scientists more research is needed, for many of those at the forefront of the field, the weight of evidence already amassed is overwhelming and fully justifies taking action. "The problem is how much should we study before we protect," asks Professor Ana Soto. "What is our duty as scientists? This issue has already hijacked our lives. How much more evidence does society need?"[23]

One of the problems with gathering the evidence is that without concerted government action it can be very difficult to tackle key questions for human health quickly. For example, Professor John Sumpter argues a great deal more could be done to identify potential routes of exposure. "I really would like to get some basic figures on human exposure to chemicals," he says. "It is entirely possible that we have missed some chemicals. Up until six months ago we missed phthalates and many of the discoveries have been made by accident. But surely if we got the major chemical manufacturers together with people who formulate the products which we use, it should be possible to draw up what you might call a 'hit list' which would allow us to assess to which chemicals we are most likely to be exposed? Senior scientists in the chemical industries could simply be asked to draw up a list of the top 100 chemicals to which there is greatest human exposure. These lists could then be compared to compile a 'hit list.' Once we had such a list, then we could test these chemicals for their ability to disrupt the endocrine system. We could test the top 100 in six months. We might find a group of chemicals which hasn't crossed anyone's mind yet."

He points out, however, that although this could be carried out very cheaply, it requires government support and pressure to ensure industry participation. It would appear even that such basic practical measures as this, to assess human exposures, are not being done. This could be solved, he says, if there was an independent body to bridge the gap between academic scientists and industry, which aimed to tackle the important questions for human health in a systematic way.

Sumpter's team at Brunel and Soto and Sonnenschein's group at Tufts are being increasingly approached by industry, who are racing to test substitute chemicals to screen them for hormonal activity. But all the companies who are ahead in this game are American or Japanese. "It is not in industry's best interests to be left behind," warns Sumpter. "Industry may lose out in the long term if governments are overprotective. Government officials are in a difficult position of calculating what information to release. But I think the British government still errs too much on the side of caution; it is too secretive and that is not in the best interests of the consumer or industry. . . . Suppose, for instance, the British government said, 'We're presently investigating whether we should allow alkylphenols to be used at all.' It just so happens they are investigating this, but very quietly. Now if that information was on some kind of public register, the companies that make alkylphenols could effectively see the writing on the wall, and would have time to prepare alternatives. Consumers would be applying a bit more pressure and it would encourage these companies to start to move. That pressure doesn't exist in Britain." Once a company finds a safe alternative, it is likely to move ahead in the market quickly. At present most companies working on alternatives are American and business opportunities are being lost in Britain, Sumpter argues.[24]

However, until we are sure we have safer substitutes, he believes it is difficult to move ahead too fast. "Yes, we should act without definitive proof of cause and effect," he says, "if we know we can make a change without introducing something worse. We don't want to replace something that is estrogenic with something that is carcinogenic. We have to be reasonably convinced that we can move to safer substitutes, and then, yes, we should do this . . . history cautions us about assuming the picture we have now is the right picture. As techniques improve and we find out more, problems appear at lower levels of exposure. We really should be concerned now. History warns us to be careful."

This is a widely held view. "I think in the end we will have to take precautions based on *evidence* rather than *proof*," agrees Professor Niels Skakkebaek. "It is the same with smoking. There is no proof that smoking causes lung cancer. But there is certainly a lot of evidence."

"In the case of cigarette smoking and lung cancer, we waited quite a long time before taking any action," says Dr. Devra Lee Davis. "We waited while scientists had debates about the quality of the data, and while industry, of course, weighed in and assured us that there was no problem; smoking was really associated with other things and didn't really cause poor health. We now know the tragedy of this. Some people have projected that there might be ten million deaths by the year 2000 from tobacco smoke. Now what if we had listened to the scientists who urged that we take action earlier? We would have all been better off."

Several scientists are so concerned at the "weight of evidence" that they argue our entire approach to regulating the industries involved now needs to be rethought. A legislative framework needs to be put in place which will give humans and wildlife much greater protection. Dr. Theo Colborn and her colleagues argue that the contamination of the globe with biologically active and persistent chemicals has sufficiently worrying implications for human health that it requires an international treaty to implement global controls. In much the same way that governments united to deal with the threats posed by ozone-depleting chemicals such as chlorofluoro-carbons in the Montreal Protocol of 1987, a similar international treaty is needed "to halt the use and dispersal of biologically active persistent compounds such as PCBs, DDT and lindane."[25] International protocols need to be developed which would "phase out the production and use of these compounds world-wide and provide institutional and financial support for their containment, retrieval and clean-up."[26]

In addition to this, Dr. Theo Colborn argues individual countries can introduce legislative measures to give greater protection. "In essence what we have to do now is make sure that we revisit every piece of legislation that is coming up for reauthorization, to make sure that we include not only cancer as a risk element, but that we include these trans-generational health effects, the effects on the developing hormone, immune and nervous systems, which are all linked. When we do that, then we can strengthen regulators' roles in controlling these chemicals."[27] Legislation needs to be redesigned to take account of accumulative exposures, interactions between exposures, and to protect those most vulnerable: the unborn.

"We have too much at stake here, all of us," says Dr. Devra Lee Davis. "What we have to do is pay better heed to experimental studies, to wildlife studies, and now to the growing body of human evidence that we have, all of which point in the same direction. All these studies suggest that there are widespread exposures to some environmental chemicals that disrupt hormones and that some of these could be behind the increases in breast cancer, the increases in testicular cancer, perhaps the increases in prostate cancer, and I would argue we cannot afford to ignore this evidence. We cannot afford to run the risk that, by ignoring this, we may take a course of action that will endanger the ability of the species to persevere."

"The findings demand that there be policy changes," advises Professor Louis Guillette. "If it's released into the environment, and you can eat it, breathe it or drink it, then it should be tested, not only for whether it causes an effect in the adult, but whether it affects a developing child as well. . . . The chemical companies have to realize at this point that many of the compounds they are releasing are having an effect on the embryo, the unborn. Is this right? No, I think it is absolutely, ethically, wrong."

"Basically we are talking about a matter of survival, not only of wildlife, but of humans as well," warns Dr. Theo Colborn. "You can reach a point of no return, where it is too late and there is nothing you can do. We're going to have to decide how long we want to wait, and how much more evidence we want to collect before we do something. This could determine how successful we are in turning things around."[28]

"At the beginning of the scientific revolution, science was something you did to understand nature, not to transform it," says Professor Ana Soto. "But now we think we know enough to go and modify nature because we are going to improve on it or get some other advantage. It is not that this shouldn't be done. It is just that this shouldn't be done so cavalierly. We imagine we have the upper hand and that is the arrogance. At the moment we are doing this as though what we do not know doesn't exist, and that is the sin."

Perhaps by now we should have learned the lesson. Nature seemed so simple: the flower in the garden, the warmth of the sun, the familiar pattern of the seasons. Yet time and time again, science has revealed an extraordinary complexity that cannot be readily encompassed by the human mind. Who would have thought that the chemicals that transformed our lives, those bright new discoveries that seemed to bring us closer to some tantalizing golden age, could be literally transforming us, biologically, in a mosaic of subtle adverse effects.

Yet, some highly regarded scientists have produced a growing body of evidence, which surely cannot be swept under the carpet by big business, governments or our own complacency. Although it is a complex problem, their evidence shows that there is an enormous contamination of the environment and possibly a dangerous one. We now know this contamination is in the soil, the rain, the oceans, the food we eat, the water we drink, the air we breathe. Nowhere can you go to escape its touch.

Postscript

Professor John Sumpter took a special interest in keeping up to date with the scientific literature, but as the information on different compounds accumulated, he began to feel more and more uneasy. It wasn't just concern at the random nature of the discoveries and the uncertainties over whether key routes of exposure had been identified. It was something more fundamental: why was it, he wondered, that we had been using some of these compounds for years and yet basic information about their biological activity had eluded us? Was it possible some of this had been known before?

He recollected that a chemist, Charles Dodds, had written to *Nature* in the 1930s about his discovery of the first *synthetic* estrogen, the notorious diethylstilboestrol, or DES. It was well known that Dodds had arrived at his discovery quite systematically by testing compounds with a similar structure to estrogen. John Sumpter began to wonder if, in the course of his experiments, Dodds could have tested some of these other estrogenic chemicals as well.

Dodds was working on families of chemicals that appeared similar to estrogen, chemicals known as the stilbenes and styrenes. Working from labs at the Courtauld Institute of Biochemistry at Middlesex Hospital, with a team from the Cancer Hospital in Chelsea, at the time their work caused quite a stir. But what was in his letters to *Nature*?

Tracking down the letters was not as straightforward as it might seem, since most computer databases do not go back to the 1930s. Ed Routledge at Brunel University decided to take this further. Eventually the letters were traced and faxed through to the biochemistry labs at Brunel. Some of the terminology used by chemists in the 1930s was different from the chemical names of today, so it took a little while to work out the meaning. Gradually, however, a disturbing picture began to emerge.

As early as January 1933, having studied the chemical properties of ovarian hormones, Dodds had foreseen the possibilities: "It seems likely that a *whole group of substances* of related chemical constitution will be found to have estrus-exciting properties [to act like an estrogen]," he wrote.[29] He had then started to search for these counterfeit hormones, realizing they may have some value in medical treatments, checking out all the likely chemical candidates in his laboratories. In the next month's edition of *Nature*, Dodds was hot on the trail, warning that "because cell proliferation which characterizes the estrus state is in some respects reminiscent of the early stages of a malignant growth, we have sought a correlation between substances having estrogenic action and those having *carcinogenic* properties." He then explained how he found that two potent carcinogens had estrogenic activity as well, a result which he found "striking" since it was thought unusual that two types of biological activity should be shown by one and the same compound.[30] At the time, he thought a common feature of the molecules termed the "phenanthrene nucleus" was responsible for the estrogenic properties and he started to investigate further. As he did so, he soon found that molecules which didn't have this particular feature could also act like estrogens. This was the breakthrough.

In 1936 he reported that a completely different class of compounds had estrogenic activity: he called them: the "diphenyls." In fact he published an entire chart of diphenyls which could produce estrus or heat in a rat.[31] Many of the compounds which have subsequently been shown to be a problem have two phenyl groups. Known today as the *biphenyls,* they include DES, bisphenol A and the estrogenic PCBs. Couldn't this have been foreseen from Dodds's letter to *Nature* in 1936, thought Sumpter. Seven months later Dodds reported another discovery. He realized that you didn't need two phenyl rings: "so simple a compound . . . containing only one benzene ring" could produce the estrogenic response. He then proceeded to characterize the estrogenic activity of certain *alkylphenols.*[32]

John Sumpter felt almost like a time traveler, viewing down the decades through the war years and the depression to the tiny office that he knew Dodds had had, marveling at the man's sheer genius. In the 1930s, the kind of equipment and resources available to him would have been more reminiscent of the last century. Yet despite the limitations, this one man had foreseen that this chemistry was going to be a mixed blessing. Wrapped up in Charles Dodds's scientific letters from the thirties were all the clues needed to unravel the story, Sumpter thought. There they were: the biphenylic compounds, such as DES and bisphenol A; some of the alkylphenols; and many other related compounds. Some were the very same classes of chemicals whose estrogenic activity was rediscovered by accident in the late eighties when they were contaminating Professor Soto's experiments in Boston and Professor Feldman's research in California. So many of the compounds which now contaminate the environment and ourselves, once created, are not easily destroyed. Dodds, of course, could not have anticipated this. Yet there it was, the essence of the story set out in black and white, as long ago as the 1930s.

What happened to this information in the intervening years remains a mystery. Quite who knew what and when, and whether industrialists were aware of Dodds's research when they began to manufacture products with these and related chemicals, raises another set of questions yet to be answered. Sumpter carefully filed away the old letters to *Nature.* "Charles Dodds," he thought, "really was quite somebody."

REFERENCES

Foreword
1. P. D. James, 1994. *The Children of Men*. Penguin.
2. Memo dated 8 February 1996 from the Endocrine Issues Task Group of the Chemical Manufacturers Association and Associated Trade Organizations to the Environment, Health, Safety and Operations Committee.

Prologue
1. Interview with Professor Louis Guillette for *Horizon*, 1993.
2. As Ref 1.
3. L. J. Guillette Jr., T. S. Gross, . . . , H. F. Percival and A. R. Woodward, 1994. *Environmental Health Perspectives*, 102: 680–687.
4. HMSO, 1992. *Effects of Trace Organics on Fish*, FR/D 0008.
5. Unpublished study. Personal communication with the author. March 1993: Roland Billard, Musée Nationale d'Histoire Naturelle, Paris.
6. C. F. Facemire, T. S. Gross and L. J. Guillette Jr., 1995. Reproductive impairment in the Florida panther: Nature or nurture? *Environmental Health Perspectives*, 103: 79–85.
7. J. F. Leatherland, 1992. Endocrine and reproductive function in Great Lakes salmon. In: *Chemically Induced Alterations in Sexual Functioning and Development* (ed. T. Colborn and C. Clements). Princeton Scientific Publishing.
8. As Ref 1.
9. S. Irvine, E. Cawood, D. Richardson, E. MacDonald and J. Aitken, 1996. Evidence of deteriorating semen quality in the United Kingdom *BMJ*, 312: 467–471.
10. Quotes in this section are from interviews with Dr. Stewart Irvine in January 1996 for *Horizon* and August 1996.

11. Memo dated 8 February 1996 from the Endocrine Issues Task Group of the Chemical Manufacturers Association and Associated Trade Organizations to the Environment, Health, Safety and Operations Committee.
12. As Ref. 11.
13. Foreword by Vice-President Al Gore to *Our Stolen Future* by Theo Colburn, Dianne Dumanoski and John Peterson Myers, Little, Brown, 1996.

Chapter One: Early Warnings

1. Quotes used in this chapter are from interviews with Niels Skakkebaek and colleagues in July 1993 and August 1996.
2. N. E. Skakkebaek, 1972. Possible carcinoma-*in-situ* of the testis. *Lancet,* September: 516–517.
3. A. Osterline, 1986. Diverging trends in incidence and mortality of testicular cancer in Denmark 1943–1982. *British Journal of Cancer,* 53: 501–505.
4. P. Boyle, S. N. Kaye and A. G. Robertson, 1987. Changes in testicular cancer in Scotland. *European Journal of Cancer and Clinical Oncology,* 23: 827–830. And: L. M. Brown, L. M. Pottern and J. T. Flannery, 1986. Testicular cancer in the US: Trends in incidence and mortality. *International Journal of Epidemiology,* 15: 164–170.
5. A. Giwercman and N. E. Skakkebaek, 1992. Editorial—The human testis: An organ at risk? *International Journal of Andrology,* 15: 373–375.
6. "Adrian Hill": Name changed to protect identity of the patient.
7. Reported in the Novo Nordisk Foundation, 1992, where Professor Niels Skakkebaek was nominated for the 1993 award.
8. C. Chilvers, M. C. Pike and M. E. J. Wadworth, 1984. Apparent doubling of frequency of undescended testes in England and Wales in 1962–81. *Lancet* ii: 330–332.
9. D. M. Campbell, J. A. Webb and T. B. Hargreave, 1987. Cryptorchidism in Scotland. *British Medical Journal.*
10. P. Matlai and V. Beral, 1985. Trends in congenital malformations of the external genitalia. Letter to *Lancet* i: 108.
11. M. B. Jackson, 1988. The epidemiology of cryptorchidism. *Hormone Research,* 30: 153–156.
12. As Ref. 5.
13. For further details see: Editorial—An increasing incidence of cryptorchidism and hypospadias. *Lancet:* i 1311. And: A. Giwercman *et al.,*1992. The human testis, an organ at risk? *Int. J. Androl.,* 15: 373–375.
14. C. M. K. Nelson and R. G. Bunge, 1973. Semen analysis: Evidence

for changing parameters of male fertility potential. *Fertility and Sterility*, 25: 503–507.

15. R. S. Hotchkiss and P. Grenley, 1938. Semen analysis of two hundred fertile men. *Am. J. Med. Sci.*, 196: 362. And: E. J. Farris, 1949. The number of motile spermatozoa as an index of the fertility of the man. *Journal of Urology*, 61: 1099.

16. J. Macleod and R. Z. Gold, 1951. The male factor in fertility and infertility. *Journal of Urology*, 66: 436.

17. As Ref. 14.

18. K. D. Smith and E. Steinberger, 1977. What is oligospermia? In: P. Troen and H. R. Nankin (eds.). Raven Press, p. 489. And: Z. Zukerman, K. D. Smith and E. Steinberger. Frequency distribution of sperm counts in fertile and infertile males. *Fertility and Sterility*, 28: 1310.

19. J. Macleod and Y. Wang, 1979. Male fertility potential. *Fertility and Sterility*, 31: 103–116.

20. F. Leto and F. J. Frensilli, 1981. Changing parameters of donor semen. *Fertility and Sterility*, 36: 766–770.

21. E. Carlsen, A. Giwercman, N. Keiding and N. Skakkebaek, 1992. Evidence for decreasing semen quality during the past 50 years. *BMJ*, 305: 609–612.

22. E. Greenhall and M. Vessey, 1990. The prevalence of subfertility: A review of the current confusion and a report of two new studies. *Fertility and Sterility*, 54: 978. And: W. Mosher and M. Pratt, 1990. Fecundity and infertility in the US 1965–88. No. 192. Hyattsville. MD: National Center for Health Statistics.

23. Interview with McGowan family for *Horizon*, 1993.

24. Personal communication with the author, 2 July 1996.

Chapter Two: The Paradox

1. S. Cooke, J. Porcelli and R. A. Hess, 1992. Induction of increased testis growth and sperm production in adult rats by neonatal administration of the goitrogen propylthiouracil (PTU): The critical period. *Biol. Reprod.*, 46: 146–154.

2. *Horizon* interview with Dr. Richard Sharpe, 1993.

3. As Ref. 2.

4. S. Hirobe, W. W. He, M. M. Lee and P. K. Donahue, 1992. Müllerian inhibiting substance messenger RNA expression in granulosa and Sertoli cells coincides with their mitotic activity. *Endocrinology*, 131: 854–862.

5. R. T. Frank, 1922. The ovary and the endocrinologist. *JAMA*, 78: 181–185.

6. E. Allen and E. A. Doisy, 1923. An ovarian hormone: Preliminary report of its localisation, extraction and partial purification and action in test animals. *JAMA*, 81: 819–821.

7. Personal communication with the author, 29 July 1996.

8. E. C. Dodds, L. Goldberg, W. Lawson and R. Robinson, 1938. Oestrogenic activity of alkylated stilboestrols. Letter to *Nature*, 141: 247. And: E. C. Dodds *et al.*, 1938. Letter to *Nature*, 142: 34.

9. As Ref. 7.

10. Anita Direcks and Ellen 't Hoen, 1986. DES the crime continues. In: *Adverse Effects, Women and the Pharmaceutical Industry* (ed. K. McDonnell). International Organization of Consumer Unions.

11. As Ref. 10.

12. Lacassagne, 1938. *Comptes rendus biol. (Paris)*, 129: 641.

13. As Ref. 10.

14. K. E. McMartin, A. Kennedy, P. Greenspan and J. Yam, 1978. DES: A review of its toxicity and use as a growth promotant in food producing animals. *J. Environ. Path. Toxicol.*, 1: 279–313.

15. W. J. Dieckmann, M. D. Davis and R. E. Pottinger, 1953. Does the administration of DES during pregnancy have any therapeutic value? *Am. J. Obst. Gynec.*, 66: 1062–1081.

16. As Ref. 10.

17. M. P. Vessey, 1989. Epidemiological studies of the effects of DES. In: *Perinatal and Multigenerational Carcinogenesis* (ed. N. P. Napalkov, J. M. Rice and L. Tomatis). International Agency for Research on Cancer, Lyon.

18. O. W. Smith, 1948. DES in the prevention and treatment of complications of pregnancy. *Am. J. Os. Gyn.*, 56: 821–835.

19. As Ref. 17.

20. As Ref. 15.

21. A. L. Herbst and R. E. Scully, 1970. Adenocarcinoma of the vagina in adolescence. *Cancer*, 25: 745–757.

22. As Ref. 17.

23. A. L. Herbst *et al.*, 1971. *New England Journal of Medicine*, 284: 878–881.

24. J. Rotmensch, K. Frey and A. L. Herbst, 1988. Effects on female offspring and mothers after exposure to DES. In: *Toxicity of Hormones in Perinatal Life* (ed. T. Miri and H. Nagasawa). CRC Press.

25. *Clear Cell Cancer*, a resource guide for DES-exposed daughters and

their families. By the National Cancer Institute, National Institute of Child Health and Development, and National Institutes of Health.

26. W. B. Gill *et al.*, 1988. Effects on human males of *in utero* exposure to exogenous sex hormones. In: *Toxicity of Hormones in Perinatal Life* (ed. T. Miri and H. Nagasawa). CRC Press.

Chapter Three: A Sea of Estrogens

1. E. C. Dodds, L. Goldberg and W. Lawson, 1938. Oestrogenic activity of esters of diethyl stilboestrol. Letter to *Nature*, 142: 211–212.

2. A. L. Herbst and R. E. Scully. 1970. Adenocarcinoma of the vagina in adolescence. *Cancer*, 25: 745–757.

3. J. A. McLachlan, R. Newbold and B. Bullock, 1975. Reproductive tract lesions in male mice exposed prenatally to DES. *Science*, 190: 991–992.

4. R. Santti, R. Newbold, S. Makela, L. Pylkkanen and J. A. Mc-Lachlan, 1994. Developmental oestrogenization and prostatic neoplasia. *Prostate*, 24: 67–78. And: L. Pylkkanen, S. Makela, E. Valve, P. Harkonen, S. Toikkanen and R. Santti, 1993. Prostatic dysplasia associated with increased expression of c-myc in neonatally oestrogenized mice. *J. Urol.*, 149: 1593–1601.

5. R. Newbold, B. T. Pentecost, S. Yamashita, C. Teng and J. A. McLachlan, 1989. Female gene expression in the seminal vesicle of mice after prenatal exposure to DES. *Endocrinology*, 124: 2568–2575.

6. *Pandora's Box: Goodbye Mrs Ant*, produced by Adam Curtis, BBC 2, 2 July 1992.

7. *Encyclopaedia Britannica*, 1991.

8. R. Sharpe, 1995. Another DDT connection. *Nature*, 375: 538–539.

9. As Ref. 6.

10. As Ref. 6.

11. As Ref. 6.

12. H. Burlington and V. F. Lindeman, 1950. Effect of DDT on testes and secondary sex characteristics of white Leghorn cockerels. *Proceedings of the Society for Experimental Biology and Medicine*, 74: 48–51.

13. R. Kuntzman, 1969. Drugs and enzyme induction. *Ann. Rev. Pharm.*, 9: 21–36.

14. D. B. Peakall, 1967. Pesticide induced enzyme breakdown of steroids in birds. *Nature*, 216: 505–506.

15. Rachel Carson, 1962. *Silent Spring*. Fawcett Publications, p. 25.

16. As Ref. 6.

17. Rachel Carson, 1962. *Silent Spring*. Fawcett Publications, pp. 13–15.

18. L. J. Guillette, 1995. Endocrine disrupting environmental contaminants and developmental abnormalities in embryos. *Human and Ecological Risk Assessment*, 1: 25–36.

19. As Ref. 6.

20. Gordon Rattray Taylor, 1970. *The Doomsday Book*. Thames & Hudson, p. 131.

21. As Ref. 6.

22. As Ref. 8.

23. A. A. Jensen, 1991. Levels and trends of environmental chemicals in human milk. In: *Chemical Contaminants in Human Milk* (ed. A. A. Jensen and S. A. Slorach). CRC Press.

24. Gordon Rattray Taylor, 1970. *The Doomsday Book*. Thames & Hudson, p. 128.

25. A. V. Holden and K. Marsden, 1967. Organochlorine pesticides in seals and porpoises. *Nature*, 216: 1275–1276.

26. Congressional Record, Senate, 1 May 1969, *Studies of Pesticides*, S4412–17, NCI.

27. J. L. Radomski, W. B. Deichmann and E. E. Clizer, 1968. Pesticide concentrations in the liver, brain, and adipose tissue of terminal hospital patients. *Toxicology*, 6: 209–220.

28. As Ref. 8.

29. Theo Colborn, Dianne Dumanoski and John Peterson Myers, 1996. *Our Stolen Future*. Little, Brown, p. 138.

30. Pan-American Health Organization, 1992. *Status of Malaria Programmes in the Americas*, 40th Report. WHO, Washington DC.

31. J. McLachlan *et al.*, 1980. *Estrogens in the Environment*. Elsevier.

32. Guzelian, 1976. 14 workers exposed to pesticide kepone are probably sterile. *Occupational Health and Safety Letter*, 8 March.

33. W. L. Duax and C. M. Weeks, 1980. Molecular basis of oestrogenicity: X-ray crystallographic studies. In: *Estrogens in Environment* (ed. J. McLachlan *et al.*). Elsevier.

34. P. J. H. Reijnders, 1986. Evidence PCBs have reproductive effects. *Nature*, 324: 456; C. F. Mason *et al.*, 1986. *Bull. Environ. Contam. Toxic.*, 36: 656.

35. R. Riseborough, 1970. More letters in the wind. *Environment*, 12: 16–27.

36. As Ref. 35.

37. As Ref. 35.

38. Søren Jensen, 1966. A report of a new chemical hazard. *New Scientist*, 32: 612.

39. As Ref. 38.
40. As Ref. 35.
41. As Ref. 35.
42. As Ref. 35.
43. As Ref. 35.
44. J. T. Borlakoglu and R. R. Dils, 1991. PCBs in human tissues. *Chemistry in Britain,* September, pp. 815–818.
45. D. J. Gregor, A. J. Peters, N. Jones and C. Spencer, 1995. The historical residue trend of PCBs in the Agassiz Ice Cap, Ellesmere Island, Canada. *Sci. Tot. Environ.,* 160/161: 117–126.
46. K. C. Jones, V. Burnett and K. S. Waterhouse, 1991. PCBs in the environment. *Chemistry in Britain,* May, pp. 435–438.
47. Allan Jensen, 1987. PCBs, PCDDs and PCDFs in human milk, blood and adipose tissue. *Sci. Tot. Environ.,* 64: 259–293.
48. As Ref. 47.
49. As Ref. 46.
50. As Ref. 44.
51. K. Korach, P. Sarver, J. McLachlan and J. McKinny, 1987. Estrogen receptor binding activity of PCBs: Conformationally restricted structural probes. *Molecular Pharmacology,* 33: 120–126.
52. D. Crews *et al.,* 1994. Temperature dependent sex determination in reptiles. *Developmental Genetics,* 15: 297–312.
53. D. P. Rall and J. McLachlan, 1980. Potential for exposure to estrogens in the environment. In: *Estrogens in the Environment* (ed. J. McLachlan *et al.*). Elsevier.

Chapter Four: First Suspects

1. Interview for *Horizon,* 1993.
2. The quotes from Richard Sharpe in this chapter are mostly from interviews in May 1993; additional material on phyto-estrogens recorded in January and August 1996.
3. R. Santti, R. Newbold and J. A. McLachlan, 1990. Developmental oestrogenization and prostatic neoplasia—Editorial. *Int. J. Androl.,* 13: 77–80.
4. From interviews for *Horizon,* "Fast Life in the Food Chain," 18 April 1992.
5. H. Adlercreutz, 1987. Diet, breast cancer, and sex hormone metabolism. *Annals New York Academy of Sciences,* 281–291.
6. H. W. Bennetts, E. J. Underwood and F. L. Shier, 1946. A specific

breeding problem of sheep on subterranean clover pastures in Western Australia. *Australian Veterinary Journal*, 2–13.

7. As Ref. 6.
8. K. Verdeal and D. Ryan, 1979. Naturally occurring estrogens in plant foodstuffs. *Journal of Food Protection*, 42: 577–583.
9. R. B. Bradbury and D. E. White, 1954. Estrogens and related substances in plants. *Vitamins and Hormones*, 12: 207–233.
10. As Ref. 9.
11. As Ref. 8.
12. T. B. Clarkson, M. S. Anthony and C. L. Hughes, 1995. Estrogentic soybean isoflavones and chronic disease: Risks and benefits. *TEM*, 6: 11–15.
13. As Ref. 12.

Chapter Five: Opening Pandora's Box

1. T. Colborn and C. Clement (eds.), 1992. *Chemically Induced Alterations in Sexual and Functional Development: The Wildlife-Human Connection.* Princeton Scientific Publishing.
2. Quotes from Theo Colborn in this chapter were recorded for BBC *Horizon* in August 1993.
3. For further details see: Theo Colborn, Dianne Dumanoski and John Peterson Myers, 1996. *Our Stolen Future.* Little, Brown, Chapter 2.
4. For further details see: G. R. Taylor, 1970, *The Doomsday Book.* Thames & Hudson.
5. R. L. Carson, 1962. *Silent Spring.* Fawcett Publications, Chapter 17.
6. As Ref. 3.
7. As Ref. 3.
8. As Ref. 2.
9. G. Fox, A. Gilman, D. Peakall and F. Anderka, 1978. Behavioural abnormalities of nesting Lake Ontario herring gulls. *Journal of Wildlife Management*, 42: 477–483.
10. M. D. Fry and C. K. Toone, 1981. DDT induced feminization of gull embryos. *Science*, 213: 922–924.
11. As Ref. 3.
12. Interview with Frederick vom Saal on 6 September 1996. And: F. vom Saal, 1984. The intrauterine position phenomenon: Effects on physiology, aggressive behavior and population dynamics in house mice. In: *Biological Perspectives in Aggression* (ed. K. Flannelly, R. Blanchard and D. Blanchard), Liss. And: F. vom Saal, 1989. Sexual differentiation in litter bearing mammals: Influence of sex of adjacent

fetuses *in utero. Journal of Animal Science,* 67: 1824–1840.

13. Interview with F. vom Saal on 6 September 1996.
14. As Ref. 13.
15. As Ref. 2.
16. As Ref. 2.
17. As Ref. 2.
18. As Ref. 2.
19. As Ref. 13.
20. Interview with John McLachlan on 20 July 1996.
21. Interview with William Davis on 5 September 1996.
22. As Ref. 20.
23. H. Bern, The fragile fetus. Chapter 2 in Ref. 1.
24. As Ref. 23.
25. J. A. McLachlan, R. Newbold, C. T. Teng and K. Korach, Environmental oestrogens: Orphan Receptors and genetic imprinting. Chapter 5 in Ref. 1.
26. As Ref. 25.
27. As Ref. 12.
28. W. P. Davis and S. A. Bortone. Effects of kraft mill effluent on the sexuality of fishes: An environmental early warning? Chapter 6 in Ref. 1.
29. P. J. Reijnders and S. M. Brasseur. Xenobiotic induced hormonal and associated developmental disorders in marine organisms and related effects in humans: An overview. Chapter 9 in Ref. 1.
30. P. Reijnders, 1986. Reproductive failure in common seals feeding off fish from polluted coastal waters. Letter to *Nature,* 324: 456–457.
31. As Ref. 29.
32. G. A. Fox. Epidemiological and pathobiological evidence of contaminant induced alternations in sexual development in free living wildlife. Chapter 8 in Ref. 1.
33. As Ref. 32.
34. Interview with John Leatherland on 5 September 1996.
35. As Ref. 12.
36. As Ref. 20.
37. As Ref. 2.
38. As Ref. 1.
39. *New York Times,* 21–26 March 1993.
40. Quotes from Dr. Richard Sharpe in this chapter were recorded for BBC *Horizon* in June 1993.

Chapter Six: Secret British Experiments

1. Quotes from Professor Sumpter in this chapter were recorded in June 1993 and July 1996.
2. Quotes from Dr. Charles Tyler were recorded in July 1996.
3. Quotes from Dr. Peter Matthiessen were recorded in August 1993.
4. Effects of trace organics on fish, FR/D 0008. HMSO, October 1992. Foundation for Water Research.
5. Quotes from Professor Guillette and Dr. Gross were recorded in July 1993.
6. For further details see: L. J. Guillette Jr., D. A. Crain. A. A. Rooney and D. B. Pickford, 1995. Organisation versus activation: The role of endocrine disrupting contaminants during embryonic development in wildlife. *Environmental Health Perspectives,* 103: 157–163. And: L. J. Guillette Jr., D. B. Pickford, D. A. Crain, A. A. Rooney and F. Percival. 1996. Reduction in penis size and plasma testosterone concentrations in juvenile alligators living in a contaminated environment. *Gerard and Comparative Endocrinology,* 101: 32–42.
7. L. J. Guillette Jr., T. S. Gross, D. A. Gross, A. A. Rooney and F. Percival, 1995. Gonadal steroidogenesis *in vitro* from juvenile alligators obtained from contaminated or control lakes. *Environmental Health Perspectives,* 103: 31–35.
8. G. H. Heinz, F. Percival and M. L. Jennings, 1991. Contaminants in American alligator eggs from Lakes Apopka, Griffin and Okeechobee. Florida. *Environ. Monit. Assess.,* 16: 277–285. And: T. S. Gross and L. J. Guillette Jr., unpublished data.
9. As Ref. 5.

Chapter Seven: The Increasing Cast of Culprits

1. Quotes from Professor Ana Soto in this chapter were recorded in August 1993 and September 1996.
2. A. Soto, H. Justicia, J. Wray and C. Sonnenschein, 1991. *p*-nonylphenol: An estrogenic xenobiotic released from "modified" polystyrene. *Environmental Health Perspectives,* 92: 167–173.
3. Quotes from Professor Sumpter in this section were recorded in 1993.
4. National Rivers Authority, 1991. Foaming in rivers: an initial assessment of the problem in the UK. Project Report 226/2/Y.
5. Quotes from Dr. Peter Matthiessen in this section were recorded in 1993.
6. BBC *Horizon,* "Assault on the Male." First transmitted October 1993.

7. L. B. Clark, R. T. Rosen, T. G. Hartman, J. B. Louis, . . . and J. D. Rosen, 1992. Determination of alkylphenol ethoxylates and their acetic acid derivatives in drinking water by particle beam liquid chromatography/mass spectrometry. *International Journal of Analytical Chemistry,* 47: 167–180.

8. Quotes from Richard Sharpe in this section were recorded in 1993.

9. Quotes and details of the incident are from an interview with Professor David Feldman, 12 September 1996.

10. D. Feldman, L. G. Tokes, P. A. Stathis, . . . and D. Harvey, 1984. Identification of 17β-oestradiol as the estrogenic substance in *Saccharomyces cerevisiae. Proc Natl. Acad. Sci. USA,* 81: 4722–4726.

11. A. V. Krishnan, P. Stathis, S. F. Permth, L. Tokes and D. Feldman, 1993. Bisphenol A: An estrogenic substance is released from polycarbonate flasks during autoclaving. *Endocrinology,* 132: 2279–2286.

12. As Ref. 11.

13. As Ref. 9.

14. Quotes from interview with Dr. Susan Jobling on 6 August 1996.

15. D. A. Murature, S. Y. Tang, G. Steinhardt and R. C. Dougherty, 1987. Phthalate esters and semen quality parameters. *Biomed. Environ. Mass Spectrom.,* 13: 473–477.

16. S. Jobling, T. Reynolds, R. White, M. Parker and J. P. Sumpter, 1995. A variety of environmentally persistent chemicals, including some phthalate plasticizers, are weakly oestrogenic. *Environmental Health Perspectives,* 103: 582–587.

17. Quotes from Professor Sumpter and Dr. Sharpe in this section are from interviews in 1996.

18. J. Autian, 1973. Toxicity and health threats of phthalate esters: Review of the literature. *Environmental Health Perspectives,* 4: 3–26. And: IARC butylbenzyl phthalate. In: *Monographs of the Evaluation of the Carcinogenic Risk of Chemicals to Humans,* Vol. 29, 1982. International Agency for Research on Cancer, Lyon, pp. 193–202.

19. R. M. Sharpe, J. S. Fisher, M. M. Millar, S. Jobling and J. P. Sumpter, 1995. Gestational exposure of rats to xenooestrogens results in reduced testicular size and sperm production. *Environmental Health Perspectives,* 103: 2–9.

20. Chemical link found to low sperm counts. *Sunday Times,* 29 October 1995.

Chapter Eight: Routes of Exposure

1. Quotes from Professor John Sumpter in this chapter are from interviews in January and July 1996.
2. Interviews with Dr. Nicolas Olea on 4 January and 3 October 1996.
3. J. A. Brotons, M. F. Olea-Serrano, M. Villalobos and N. Olea, 1995. Xenooestrogens released from lacquer coatings in food cans. *Environmental Health Perspectives*, 102: 608–612.
4. Association of Plastic Manufacturers in Europe: Joint Statement of the Epoxy Resins Group of APME and European Confederation of Paint, Printing Ink and Artists Colours Manufacturers Associations and European Secretariat of Manufacturers of Light Metal Packaging, 22 November 1995.
5. Interview with Professor Carlos Sonnenschein on 4 January 1996.
6. N. Olea, R. Pulgar, P. Perez, F. Olea-Serrano, . . . , A. Soto and C. Sonnenschein, 1996. Estrogenicity of resin based composites and sealants used in dentistry. *Environmental Health Perspectives*, 104: 298–305.
7. Interviews with Professor Ana Soto in January and September 1996.
8. Interview with Dr. Sue Jobling on 6 August 1996.
9. Interview with Professor Frederick vom Saal on 6 September 1996.
10. American Dental Association Position Statement on Estrogenicity of Dental Sealants from the ADA Council of Scientific Affairs, Spring 1996.
11. As Ref. 2.
12. As Ref. 3.
13. D. Feldman and A. Krishnan, 1995. Estrogens in unexpected places: Possible implications for researchers and consumers. *Environmental Health Perspectives*, 103: 129–133.
14. US National Toxicology Program, 1991. Report T 0035 C: Final report on reproductive toxicity of di-*n*-butyl phthalate. NIEHS, North Carolina. See also Report NTP TR 458.
15. M. Sharman, W. Read, L. Castle and J. Gilbert, 1994. Levels of di-2-ethylhexyl phthalate and total esters in milk, cream, butter and cheese. *Food Additives and Contaminants*, 11: 375–385.
16. As Ref. 15.
17. As Ref. 15.
18. MAFF food surveillance information sheet. No. 60, May 1995.
19. B. D. Page and G. M. Lacroix, 1992. Studies into the transfer and migration of phthalate esters from aluminium foil paper laminates to butter and margarine. *Food Additives and Contaminants*, 8: 701–706.
20. MAFF, 1987. Survey of plasticizer levels in food contact materials and in foods. HMSO.

21. As Ref. 20.
22. As Ref. 15.
23. R. M. Sharpe, J. S. Fisher, M. M. Millar, S. Jobling and J. P. Sumpter, 1995. Gestational and lactational exposure of rats to xenooestrogens results in reduced testicular size and sperm production. *Environmental Health Perspectives,* 103: 1136–1143.
24. As Ref. 23.
25. As Ref. 23.
26. MAFF food surveillance information sheet. No. 82, March 1996.
27. G. Lyons. Phthalates in the environment. A report for WWF, 20 April 1996.
28. Interview with Gwynne Lyons, January 1996.
29. *Horizon,* "Assault on the Male" update. BBC 2, 12 February 1996.
30. The ministry and a culture of secrecy. *Daily Mail,* 27 May 1996.
31. Tainted baby milk storm. *Daily Mail,* 27 May 1996.
32. Sex change chemicals in baby milk. *Independent on Sunday,* 26 May 1996.
33. The five million ton plastic peril in food. *The Guardian,* 28 May 1996.
34. Doubts that linger over baby milk. *The Times,* 27 May 1996.
35. As Ref. 32.
36. Phthalates in infant formulae. MAFF food surveillance information sheet. No. 83, March 1996.
37. As Ref. 34.
38. Interview with Dr. Richard Sharpe on 16 August 1996.
39. How my work triggered the milk fiasco. *Daily Telegraph,* 5 June 1996.
40. As Ref. 38.
41. As Ref. 7.
42. A. M. Soto, K. L. Chung and C. Sonnenschein, 1994. The pesticides endosulfan, toxaphene and dieldrin have oestrogenic effects on human estrogen sensitive cells. *Environmental Health Perspectives,* 102: 380–383.
43. As Ref. 5.
44. A. M. Soto *et al.,* 1995. The E screen assay as a tool to identify estrogens: An update on estrogenic environmental pollutants. *Environmental Health Perspectives,* 103: 114–121.
45. As Ref. 5.
46. E. P. Laug, F. M. Kunze and C. S. Prickett, 1951. Occurrence of DDT in human milk and fat. *Arch. In. Hyg.,* 3: 245.
47. A. A. Jensen and S. A. Slorach (eds.), 1991. *Chemical Contaminants in Human Milk.* CRC Press.

48. A. A. Jensen, Transfer of chemical contaminants into human milk. Chapter 2 in Ref. 47, pp. 10, 11.

49. As Ref. 48, p. 15.

50. A. A. Jensen and S. A. Slorach, Factors affecting the levels of residues in human milk. Chapter 6 in Ref. 47, p. 200.

51. A. A. Jensen, Levels and trends of environmental chemicals in human milk. Chapter 5 in Ref. 47, p. 47.

52. As Ref. 51, p. 86.

53. Interview with Dr. Mary Wolff, Mount Sinai Medical Hospital, in January 1996.

54. Interview with Professor Raymond Dils on 13 September 1996. And: J. T. Borlakoglu, N. J. Borlak and R. R. Dils, 1990. Human milk: A source of potentially toxic and carcinogenic halosubstituted biphenyls. *Proceedings of the Nutrition Society*, 49–56. And: J. T. Borlakoglu and R. R. Dils, 1991. PCBs in human tissues. *Chemistry in Britain*, September: 815–818.

55. As Ref. 54.

56. J. C. Larsen. Toxicological implication of persistent organohalogens in mother's milk as indicated by animal experiments. Chapter 9 in Ref. 47, p. 246.

57. A. A. Jensen and S. A. Slorach. Overall assessment and evaluation. Chapter 11 in Ref. 47, p. 287.

58. As Ref. 57, p. 288.

59. S. H. Safe, 1995. Environmental and dietary estrogens and human health: Is there a problem? *Environmental Health Perspectives*, 103, 346–351.

60. S. H. Safe, 1995. Assessing the role of environmental oestrogens in human reproductive health. *Health and Environment Digest*, 8: 79–81.

61. Interview with Professor Safe recorded on 2 October 1996.

62. As Ref. 59.

63. Interview with Professor Stephen Safe, recorded on 2 October 1996.

64. Quote from untransmitted interview for *Horizon* with Chemical Industries Association, London, January 1996.

65. As Ref. 9.

66. As Ref. 38.

67. As Ref. 9.

68. As Ref. 53.

69. Interview with Dr. Charles Tyler on 6 August 1996.

70. A. A. Jensen, Transfer of chemical contaminants into human milk. Chapter 2 in Ref. 47, p. 10.

71. J. P. Sumpter and S. Jobling, 1995. Vitellogenesis as a biomarker for estrogenic contamination of the aquatic environment. *Environmental Health Perspectives,* 103: 174–177.

72. As Ref. 71.

73. L. G. Hansen, 1994. Letter to *Science,* 226, October.

74. As Ref. 42.

75. As Ref. 7.

76. As Ref. 71.

77. C. R. Tyler, S. Jobling and J. P. Sumpter, 1996. Chemicals which mimic hormones. In the press.

78. As Ref. 1.

Chapter Nine: "A Universe of Chemicals"

1. Quotes from Linda Birnbaum in this chapter are from a research interview for *Horizon* in August 1993 and a recorded interview on 24 July 1996.

2. Quotes from Dr. Earl Gray in this chapter were recorded on 30 July 1996.

3. E. J. Gray, J. S. Ostby and W. R. Kelce, 1994. Developmental effects of an environmental anti-androgen: The fungicide vinclozolin alters sex differentiation of the male rat. *Toxicology and Applied Pharmacology,* 129: 46–52.

4. As Ref. 3.

5. W. R. Kelce, C. Stone, S. Laws, E. J. Gray, J. Kemppainen and E. Wilson. Persistent DDT metabolite p,p'-DDE is a potent androgen receptor antagonist. *Nature,* 375: 581–585.

6. J. Emsley, 1994. *The Consumer's Good Chemical Guide.* Chapter 7.

7. As Ref. 6.

8. P. Bertazzi, A. Pesatori, D. Consommi, A. Tironi, M. Landi and C. Zocchetti, 1993. Cancer incidence in a population accidentally exposed to 2,3,7,8-tetrachlorodibenzo-*para*-dioxin. *Epidemiology,* 4: 398–406.

9. M. J. De Vito and L. S. Birnbaum, 1995. *Toxicology:* Special issue on Biological Mechanisms and Quantitative Risk Assessment, 102: 115–123.

10. As Ref. 9.

11. As Ref. 6.

12. Dioxins in human milk. MAFF food surveillance information sheet. No. 88, May 1996.

13. As Ref. 12.

14. As Ref. 12.
15. World Health Organization, Regional Office for Europe (1991) Summary Report. Consultation on TDIs from food of PCDDs and PCDFs. Bilthoven, Netherlands, 4–7 December, 1990, EUR/ICP/PCS 030 S 0369n publication. WHO Regional Office for Europe, Copenhagen.
16. As Ref. 12.
17. D. Barsotti and J. R. Allen, 1979. *Bull Environ. Contam. Toxicol.*, 211: 463–470.
18. C. H. Bamtomsky, 1977. *Endocrinology*, 101: 292–299.
19. R. Peterson, R. Moore, T. Mably, D. Bjerke and R. Goy, 1992. Male reproductive system ontogeny: Effects of perinatal exposure to 2,3,7.8-tetrachlorodibenzo-*p*-dioxin. In: *Chemically Induced Alterations in Sexual and Functional Development* (eds. T. Colborn and C. Clements). Princeton Scientific Publishing.
20. As Ref. 1.
21. As Ref. 19.
22. E. J. Gray and J. S. Ostby, 1995. *In utero* dioxin alters reproductive morphology and function in female rat offspring. *Toxicology and Applied Pharmacology*, 133: 285–294.
23. A. Pesatori, D. Consommi, A. Tironi, C. Zocchetti and P. Bertazzi, 1993. Cancer in a young population in a dioxin contaminated area. *International Journal of Epidemiology*, 22: 1010–1013.
24. Quotes from interview with Dr. Audrey Cummings on 22 July 1996.
25. S. E. Rier, D. C. Martin, R. E. Bowman and J. L. Becker, 1993. Endometriosis in rhesus monkeys following exposure to 2,3,7,8-tetrachlorodibenzo-*p*-dioxin. *Fund. Appl. Toxicol.*, 21: 433–441.
26. A. M. Cummings, J. L. Metcalf and L. Birnbaum, 1996. Promotion of endometriosis by 2,3,7,8-tetrachlorodibenzo-*p*-dioxin in rats and mice: Time-dose dependence and species comparison. *Toxicology and Applied Pharmacology*, 138: 131–139.
27. E. J. Gray, E. Monosson and W. Kelce. Emerging issues: The effects of endocrine disrupters on reproductive development. Chapter 4 in: *Interconnections between Human and Ecosystem Health* (ed. E. Monosson and R. T. DiGiulio). Chapman & Hall.
28. As Ref. 27.
29. Gray is referring here to the studies of Brauwer, Daly and Jacobson and others. For a full review, see: Colborn, Dumanoski and Myers (1996), *Our Stolen Future*. Little, Brown, Chapter 10.
30. T. Colborn, F. S. vom Saal and A. Soto, 1993. Developmental effects

of endocrine disrupting chemicals in wildlife and humans. *Environmental Health Perspectives*, 101: 378–383.

31. The quotes from Soto, Sumpter, Skakkebaek and Sharpe included in this section are from interviews conducted in 1996.

32. Theo Colborn, Dianne Dumanoski and John Peterson Myers, 1996. *Our Stolen Future.* Little, Brown, Chapter 12.

33. A. A. Jensen and S. A. Slorach (eds.), 1991. *Chemical Contaminants in Human Milk.* CRC Press. And: MAFF food surveillance sheet. No. 88. May 1996.

34. As Ref. 32.

35. As Ref. 32.

Chapter Ten: "Playing the Trump Card of Uncertainty"

1. E. Carlsen, A. Giwercman, N. Keiding and N. E. Skakkebaek, 1992. Evidence for decreasing quality of semen during the past 50 years. *BMJ*, 305: 609–613.

2. Quotes in this chapter with Dr. Stewart Irvine are from interviews in January 1996 and August 1996.

3. G. W. Olsen, K. M. Bodner, J. M. Ramlow, C. E. Ross and L. I. Lipshulz, 1995. Have sperm counts been reduced 50 percent in 50 years? A statistical model revisited. *Fertility and Sterility*, 63: 887–893.

4. N. Keiding, 1994. Falling sperm quality. *BMJ*, 309: 131. And: N. Keiding and N. E. Skakkebaek, 1997. On statistical evidence concerning a possibly decreasing sperm concentration. *Fertility and Sterility*. In the press.

5. W. H. James, 1980. Secular trends in reported sperm count. *Andrologia*, 12: 381–388.

6. Interview with Dr. Ewa Rajpert-De Meyts on 9 August 1996.

7. Quotes from Dr. Richard Sharpe in this chapter are from an interview in August 1996.

8. J. Auger, M. D. Kunnstmann, F. Czyglik and P. Jouannet, 1995. Decline in semen quality among fertile men in Paris during the past 20 years. *New England Journal of Medicine*, 332: 281–285.

9. As Ref. 8.

10. R. Sherins, 1995. Are semen quality and male fertility changing? *New England Journal of Medicine*, 332: 327.

11. J. de Mouzon and P. Thonneau, 1996. Declining sperm count. Letter to the editor, *BMJ*, 313: 43.

12. K. van Waeleghem, N. de Clercq, L. Vermeulen, . . . and F. Comhaire, 1996. Deterioration of sperm quality in young healthy Belgian

men. *Human Reproduction,* 11: 325–329. And: K. van Waeleghem, N. de Clercq, L. Vermeulen, . . . and F. Comhaire, 1994. Deterioration of sperm quality in young Belgian men during recent decades. Abstracts of the 10th Annual Meeting of the ESHRE, Brussels.

13. Quotes from Professor Comhaire are from an interview on 2 July 1996.

14. As Ref. 13.

15. As Ref. 13.

16. J. Ginsberg, S. Okolo, G. Prelevic and P. Hardiman, 1994. Residence in London area and sperm density. *Lancet,* 343: 230.

17. As Ref. 16.

18. As Ref. 2.

19. D. S. Irvine *et al.,* 1996. Evidence of deteriorating semen quality in the UK: Birth cohort study in 577 men in Scotland over 11 years. *BMJ,* 312: 470–471.

20. As Ref. 2.

21. S. Farrow, 1996. Results cannot be generalized. *BMJ,* 313: 43–44.

22. As Ref. 2.

23. This data is summarized in: "Silent sperm" by Lawrence Wright, in the *New Yorker,* 15 January 1996.

24. These quotes are from an interview recorded for *Horizon* in January 1996.

25. H. Fisch, J. Feldshuh, J. H. Olsen and D. H. Barad, 1996. Semen analysis of 1283 men from the US over a 25-year period: no decline in quality. *Fertility and Sterility,* 65: 1009–1015.

26. Quotes from an interview with Professor Fisch recorded on 2 October 1996.

27. Quotes from an interview with Professor Stephen Safe recorded on 2 October 1996.

28. As Ref. 7.

29. M. Vierula *et al.* High and unchanged sperm counts of Finnish men. *International Journal of Andrology,* 19: 46–51.

30. Quote from *BMJ* press release on 20 February 1996, citing study of Dr. L. Bujan, Centre Hôpital Universitaire La Grave, Toulouse.

31. As Ref. 27.

32. T. K. Jensen *et al.,* 1996. Semen quality among members of organic food associations in Zealand, Denmark. *Lancet,* June.

33. N. Olea *et al.,* 1996. Exposure to pesticides and cryptorchidism: Geographical evidence of a possible association. *Environmental Health Perspectives,* 104: 2–8.

34. E. Lynge, 1996. Time trends in semen quality: What have we learned? The Danish Cancer Society, July.

35. Interview with Professor Niels Skakkebaek on 9 August 1996.

36. H. O. Adami *et al.*, 1994. Testicular cancer in nine northern European countries. *Int. J. Cancer,* 59: 33–38.

37. S. S. Devesa *et al.*, 1995. Recent cancer trends in the United States. *Journal of the National Cancer Institute,* 87: 175–182.

38. E. Hoff Wanderas, S. Tretli and S. D. Fossa, 1995. Trends in incidence of testicular cancer in Norway 1955–1992. *European Journal of Cancer* A, 31: 2044–2048.

39. Playing the trump card of uncertainty. Editorial, *Nature,* 1996, 380: 1.

40. Interview with Professor Carlos Sonnenschein for *Horizon* in January 1996.

Chapter Eleven: The Human Price

1. Quotes in this chapter were recorded in an interview with Dr. Henrik Leffers on 9 August 1996.

2. Quotes from Professor Stephen Safe in this chapter are from an interview on 2 October 1996.

3. Quotes in this chapter were recorded in an interview with Dr. Philippa Darbre on 18 September 1996.

4. This issue is summarized in: D. L. Davis and H. L. Bradlow, 1995. Can environmental oestrogens cause breast cancer? *Scientific American,* October, p. 143 onward. And: M. S. Wolff: and P. G. Toniolo, 1995. Environmental organochlorine exposure as a potential etiologic factor in breast cancer. *Environmental Health Perspectives,* 103: 141–145.

5. This data is summarized in: N. Krieger *et al.*, 1994. Breast cancer and serum organochlorines. *Journal of the National Cancer Institute,* 86: 589–599. For more details see also: A. A. Jensen, 1988. *Drugs and Human Lactation.* Elsevier, pp. 551–573.

6. E. J. Freur and L. M. Wun, 1992. *Am. J. Epidemiol.,* 136: 1423. And cited in: A. Soto, 1996. Control of estrogen target cell proliferation, environmental estrogens and breast tumorigenesis. *Toxicology,* 5: 425–433.

7. Dr. Mary Wolff interviewed for *Horizon,* January 1996.

8. V. Beral, C. Hermon, G. Reeves and R. Peto, 1995. Sudden fall in breast cancer death rates in England and Wales, *Lancet,* 345: 1642–1643.

9. W. J. Rogan *et al.*, 1987. PCBs and DDE in human milk: Effects on

growth, morbidity and duration of lactation. *Am. J. Public Health,* 77: 1294–1297.

10. B. C. Gladen and W. J. Rogan, 1995. DDE and shortened duration of lactation in a northern Mexican town. *Am. J. Public Health,* 85: 504–509.

11. F. Y. Falck *et al.,* 1992. Pesticides and PCB residues in human breast lipids and their relation to breast cancer. *Arch. Environ. Health,* 47: 143–146.

12. M. S. Wolff *et al.,* 1993. Blood levels of organochlorine residues and the risk of breast cancer. *J. Natl. Cancer Inst.,* 85: 648–652.

13. N. Krieger *et al.,* 1994. Breast cancer and serum organochlorine: A prospective study among white, black and Asian women. *J. Natl. Cancer Inst.,* 86: 589–599.

14. As Ref. 13.

15. As Ref. 2.

16. Quotes from Professor Devra Lee Davis in this chapter are from an interview for *Horizon* in January 1996.

17. Quotes from Professor Frederick vom Saal are from an interview on 27 September 1996.

18. M. Wassermann *et al.,* 1976. Organochlorine compounds in neoplastic and adjacent apparently normal breast tissue. *Bulletin of Environmental Contamination and Toxicology,* 15: 478–484.

19. P. Pujol, S. G. Hilsenbeck, G. C. Chamness and R. M. Elledge. 1994. Rising levels of oestrogen receptor in breast cancer over two decades. *Cancer,* 74: 1601–1606.

20. E. Dewailly *et al.,* 1994. High organochlorines body burden in women with oestrogen receptor positive breast cancer. *J. Natl. Cancer Inst.,* 86: 232–235.

21. H. Mussalo Rauhamaa *et al.,* 1990. Occurrence of β-hexachlorocyclohexane. *Cancer,* 66: 2125–2128.

22. S. H. Safe, 1995. Environmental and dietary oestrogens and human health: Is there a problem? *Environmental Health Perspectives,* 103: 346–351.

23. M. S. Wolff and P. G. Toniolo, 1995. Environmental organochlorine exposure as a potential etiologic factor in breast cancer. *Environmental Health Perspectives,* 103: 141–145.

24. Further details of Professor Raymond Dils's studies on PCBs and breast milk are outlined in Chapter 8.

25. As Ref. 2.

26. Further details of Professor John McLachlan's studies of the estrogen-icity of PCBs are outlined in Chapter 3.

27. Further details of the research of Earl Gray and others on how PCBs and dioxins can interact with the Ah receptor are described in Chapter 9.

28. K. Nesaretnam, D. Corcoran, R. R. Dils and P. Darbre, 1996. 3,4,3',4'—Tetrachlorobiphenyl acts as an oestrogen *in vitro* and *in vivo*. *Molecular Endocrinology*, 10: 923–936.

29. This study is in the press.

30. Quotes in this section are from an interview with Dr. Michael Osborne for *Horizon* in January 1996.

31. D. W. Sepkovic, H. L. Bradlow, G. Ho, S. E. Hankinson, L. Gong, M. P. Osborne and J. Fishman, 1995. Estrogen metabolite rations and risk assessment of hormone related cancers. *Annals of the New York Academy of Sciences*, 768: 312–316.

32. D. L. Davis and H. L. Bradlow, 1995. Can environmental oestrogens cause breast cancer. *Scientific American*, October: 144–149.

33. H. L. Bradlow, D. L. Davis, G. Lin, D. W. Sepkovic and R. Tiwari, 1995. Effects of pesticides on the ratio of $16\alpha/2$-hydroxyestrone: A biologic marker for breast cancer risk. *Environmental Health Perspectives*, 103: 1–4.

34. As Ref. 30.

35. As Ref. 16.

36. As Ref. 32.

37. Studies showing testicular cancer may have its origins in fetal life are described in Chapter 1.

38. Chapter 2 outlines the mechanism whereby overexposure to estrogens in the womb would damage male reproductive health.

39. Chapter 3 outlines McLachlan and Santti's early work on prostate cancer, showing how it can be induced with prenatal synthetic estrogen exposure.

40. D. J. Nonneman, V. K. Ganjam, W. V. Welshons and F. S. vom Saal, 1992. Intrauterine position effects on steroid metabolism and steroid receptors of reproductive organs in male mice. *Biology of Reproduction*, 47: 723–729.

41. As Ref. 17.

42. F. S. vom Saal, 1995. Environmental estrogenic chemicals and their impact on embryonic development. *Human and Ecological Risk Assessment*, 1: 3–15.

43. As Ref. 17.

44. As Ref. 16.
45. H. A. Tilsen, J. L. Jacobson and W. Rogan, 1990. Polychlorinated biphenyls and the developing nervous system: Cross species comparisons. *Neurotoxicology and Teratology*, 12: 239–248.
46. A. Brouwer *et al.*, 1995. Functional aspects of developmental toxicity of polyhalogenated aromatic hydrocarbons in experimental animals and human infants. *European Journal of Pharmacology*.
47. As Ref. 46.
48. G. G. Fein *et al.*, 1984. Prenatal exposure to PCBs: Effects on birth size and gestational age. *Journal of Pediatrics*, 105: 315–320.
49. As Ref. 48.
50. As Ref. 48.
51. As Ref. 46.
52. As Ref. 46.
53. For more information see: H. Daly, 1992. The evaluation of behavioural changes produced by consumption of environmentally contaminated fish. In: *Malnutrition and Hazard Assessment*, Vol. 1 (ed. R. Isaacson and K. Jensen). Plenum.
54. As Ref. 46.
55. P. Ross, R. de Swart, P. Reijnders and A. Osterhaus, 1995. Contaminant related suppression of delayed type hypersensitivity and antibody responses in harbor seals fed herring from the Baltic Sea. *Environmental Health Perspectives*, 103: 162–167.
56. Consensus statement was released at a press conference in Washington on 30 May 1996. The work session on "Environmental endocrine disrupting chemicals: Neural, endocrine and behavioral effects' was held on 5 November 1995 in Erice, Sicily.
57. F. S. vom Saal *et al.*, 1995. Estrogenic pesticides: Binding relative to estradiol in MCF 7 cells and effects of exposure during fetal life on subsequent territorial behavior in male mice. *Toxicology Letters*, 77: 343–350. And: M. Hines, 1996. Surrounded by estrogens? Considerations for neurobehavioral development in human beings. Cited in Ref. 56.
58. As Ref. 56.
59. Theo Colborn, Dianne Dumanoski and John Peterson Myers, 1996. *Our Stolen Future*. Little, Brown.

Chapter Twelve: The Response

1. Quotes from interview with John McLachlan on 10 July 1996.

2. European Chemical Industry Council Environmental oestrogens and endocrine modulators, 19 January 1996.
3. Endocrine Issues Coalition Research Summary, April 1996. Further information can be obtained from the coalition members including: the American Crop Protection Association; the Chemical Manufacturers Association USA; and the Society of the Plastic Industry, USA.
4. As Ref. 3.
5. Chemical Industries Association, Smith Square. London. Industry Brief.
6. Untransmitted interview for *Horizon* with spokesperson for the Chemical Industries Association, January 1996.
7. Quotes in this chapter are from an interview with Dr. Richard Sharpe in August 1996.
8. Interview with senior representative of chemical industries in America who cannot be identified, 27 September 1996.
9. Interview with Professor John Sumpter in August 1996.
10. Memo from the Endocrine Issues Task Group of the CMA to the Environment Health and Safety and Operations Committee, 8 February 1996.
11. Agreements from the Inter-Association Meeting on Endocrine Issues, 29 September 1995, at the University Club.
12. As Ref. 10.
13. As Ref. 10.
14. As Ref. 2.
15. Theo Colborn, Dianne Dumanoski and John Peterson Myers, 1996. *Our Stolen Future*. Little, Brown.
16. J. Toppari *et al.*, 1995. Male reproductive health and environmental chemicals with estrogenic effects. Danish Environmental Protection Agency, No. 290.
17. As Ref. 16.
18. As Ref. 16.
19. Institute for Environment and Health. Environment oestrogens: Consequences to human health and wildlife, 1995.
20. As Ref. 19.
21. Not heeding one's own advice. Focus Report, Environmental Data Services, July.
22. As Ref. 9.
23. Interview with Professor Ana Soto in July 1993.
24. As Ref. 9.
25. As Ref. 15.

26. As Ref. 15.
27. Quotes from an interview with Dr. Theo Colborn for *Horizon* in August 1993.
28. As Ref. 27.
29. J. W. Cook, E. C. Dodds and C. L. Hewett, 1933. A synthetic oestrus-exciting compound. *Nature,* January: 56–57.
30. J. W. Cook and E. C. Dodds, 1933. Sex hormones and cancer producing compounds. *Nature,* February: 205–206.
31. E. C. Dodds and W. Lawson, 1936. Synthetic oestrogenic agents without the phenanthrene nucleus. *Nature,* June: 996.
32. E. C. Dodds and W. Lawson, 1937. A simple aromatic oestrogenic agent with an activity of the same order as that of oestrone. *Nature,* April: 627–628.

INDEX